LEARNING AND DOING MATHEMATICS

John Mason

Professor of Mathematical Education
The Open University

No.1 in QED's

Visions of Mathematics

Series

 Prepared originally for The Open University.
New edition produced by QED.

© The Open University 1988   © John Mason 1999   1-85853-049-0

HOW TO SOLVE IT

First. You have to *understand* the problem.	**1. Understand the problem** • *What is the unknown? What are the data? What is the condition?* • Is it possible to satisfy the condition? Is the condition sufficient to determine the unknown? Or is it insufficient? Or redundant? Or contradictory? • Draw a figure. Introduce suitable notation. • Separate the various parts of the condition. Can you write them down?
Second. Find the connection between the data and the unknown. You may be obliged to consider auxiliary problems if an immediate connection cannot be found. You should obtain eventually a *plan* of the solution.	**2. Devise a plan** • *Have you seen it before?* *Have you seen the same problem in a slightly different form?* • *Do you know a related problem?* Do you know a theorem that could be useful? • *Look at the unknown!* Try to think of a problem you know which has the same or a similar unknown. • *Here is a problem related to yours and solved before. Can you use it?* Could you use its result? Could you use its method? Should you introduce some auxiliary element in order to make its use possible? • Could you restate the problem? Could you restate it still differently? Go back to definitions. • If you cannot solve the proposed problem, first try to solve some related problem. Could you imagine a more accessible related problem? A more general problem? A more special problem? An analogous problem? Could you solve a part of the problem? Keep only a part of the condition, drop the other part; how far is the unknown then determined, how can it vary? Could you derive something useful from the data? Could you think of other data appropriate to determine the unknown? Could you change the unknown or the data, or both if necessary, so that the new unknown and the new data are nearer to each other? • Did you use all the data? Did you use the whole condition? Have you taken into account all essential notions involved in the problem?
Third. *Carry out* your plan.	**3. Carry out the plan** • Carrying out your plan of the solution *check each step*. Can you see clearly that the step is correct? Can you *prove* that it is correct?
Fourth. *Examine* the solution obtained.	**4. Look back** • Can you *check the result?* Can you check the argument? • Can you derive the result differently? Can you see it at a glance? • Can you use the result, or the method, for some other problem?

Problem Solving Strategies. Adapted from *Problem Solving in School Mathematics*, contributors G. Polya *et al.*

CONTENTS

Other books by John Mason

Questions and Prompts for Mathematical Thinking.
ATM (Derby), 1998, with Anne Watson.

Mathematics Foundation Problem Book, 1971-1996.
Centre for Mathematics Education, The Open University.

Personal Enquiry: Moving from Concern towards Research.
The Open University, 1996.

Teaching Mathematics: Action and Awareness. The Open University, 1992.

Supporting Primary Mathematics: Algebra. The Open University, 1991.

Modelling with Mathematics in Primary and Secondary Schools.
Fostering and Sustaining Mathematics Thinking through Problem Solving.
Deakin University (Geelong, Australia), 1991, with Joy Davis.

Expressing Generality
Doing & Undoing
Actions into Words
Approaching Infinity
Dealing with Decimals. The Open University, 1988.

Thinking Mathematically. Addison Wesley, 1985 (with Burton & Stacey).

Further recommended reading

G. Polya: **Mathematical Discovery.**
How to Solve It
Mathematics and Plausible Reasoning
Excellent books, full of challenging problems and good advice from the master.

Raymond Smullyan: logic books such as **"What is the Name of this Book?"**
Excellent for sharpening one's thinking.

A. Gardiner: **Discovering Mathematics: the Art of Investigation.**
One of several by this author with challenging problems.

Stephen Brown & Marion Walter:
The Art of Problem Posing.
Problem Posing: Reflections and Applications
Excellent advice on posing one's own problems.

G. Polya and others: **Problem Solving in School Mathematics.** NCTM.
A hardbound NCTM Yearbook with 22 articles including an annotated
bibliography.

See also references on pp. 60-61. All these books and *any book in print* can be obtained from QED Books (maths specialists): Tel: 0345-402275 for school and personal orders, or email *qedbooks@aol.com*

Related websites

www.nrich.maths.org.uk   www.srl.rmit.edu.au/mav/PSTC/index.html

PREFACE TO NEW EDITION

The Essence of Learning is *Doing*

Teaching fashions may change, but the essence of learning remains constant: in mathematics, *the essence of learning is doing*.

The original edition of this book, which was part of The Open University Mathematics Foundation Course M101, emphasized the central core of doing mathematics as:

- specializing (constructing particular examples to see what happens), and

- generalizing (detect a form; express it as a conjecture; then justify it through reasoned argument).

Specializing and Generalizing have subsequently entered the common language of maths educators, but I continue to be astonished at how often they are overlooked, to the detriment of thinking. This book may have its faults (it omits much in order to be approachable; it may overstress in order to be heard), but the new edition has changed little because its message remains valid despite the passage of time and changes of fashion.

A key and unusual feature of the book remains that its advice on study methods is directly related to and builds upon the fundamental processes of mathematical thinking.

Whatever the latest rhetoric (e.g. "whole class interactive teaching") *Learning and Doing* provides a still point in the storm. It emphasizes that students must *do* mathematics. Furthermore, all six strategic modes of interaction - "the 6 Ex's" - must be employed:

Expound Explore Exercise Explain Examine Express

Learning

Modern mathematics, however complex, may be traced back to simple beginnings in arithmetic. But it is much more than 'glorified arithmetic'.

Similarly, learning mathematics involves much more than simply mastering techniques for routine unproblematic 'problems'. Learning mathematics involves adopting a particular way of perceiving the world - a particular perspective.

Learning and Doing shows that the fundamentals of that perspective apply to learning other people's worked-out mathematics (ideas, theorems, techniques), as much as they apply to doing mathematics yourself. Indeed, in my experience, to learn mathematics you must do mathematics. Learning and Doing shows how this can be done.

The aim of mathematics is to make and justify conjectures, not just to learn techniques. Mathematical practice can be seen as a constant process of turning the problematic into the routine. This is done by locating relationships and structures which enable whole classes of problems to be solved - while at the same time revealing yet more problems.

Unfortunately mathematics as presented often obscures this: it ignores the struggles which alone can create an appropriate structure and perspective from which solutions are possible, and presents instead just a bland routine - the product, not the process.

Learning and Doing shows how specializing and generalizing can help you make someone else's routine into an object of study and exploration. In this way you can come to own an expression which is similar to, perhaps even identical with, what they wrote or said. Once you have constructed it yourself, you own it - and by owning it you are better able to re-construct and further develop it in the future.

The need for maximal involvement

You may feel you only want a certain amount of mathematical technique in order to pursue your scientific or technological interests. But phrasing it that way already puts you in a weak position. What you really need could be an appreciation of mathematical thinking, together with exposure to certain techniques and results. You may be tempted to minimize involvement in mathematical thinking, hoping that mastery of a few techniques will be sufficient. But this view is flawed in at least four ways:

- It is more efficient to understand than merely to internalize or automate; it is possible to train behaviour without educating awareness.

- To respond to new situations flexibly and creatively you must *own* active links and connections between different mathematical ideas - thinking for yourself rather than merely applying standard techniques to standard question-types. (Awareness needs to be educated as well as behaviour trained).

- Active participation and thought releases energy which can be harnessed productively; passive participation absorbs energy.

- The fundamental processes introduced here apply to every discipline, though in different ways. So your awareness of your own powers of specializing and generalizing can help you in whatever discipline you pursue.

Doing is not construing!

Every mathematical activity in which I participate reinforces my view that the foundation of learning mathematics lies in doing things: "doing" entails creating and manipulating confidence-inspiring entities to get a sense of what is 'going on'. These confidence-inspiring entities can be material objects, but more often they are images, diagrams and graphs, and symbols. This visualization is central to mathematical doing.

But doing in itself is far from sufficient. You may "do the exercises", even "do well" on them, but that does not mean you really understand. As one student told Sherman Stein

> *"In high school the teacher works one problem on the board and then we do twenty just like it. We don't have to* know *anything."*[1]

This may be summarized as *Doing is not Construing*.

Rather than 'doing all the odd numbered exercises' with little thought about where they are the same and where they differ, I recommend the following:

1. Do a few exercises until you have a feel for their common structure (the "meta-question")

2. Consider this meta-question, and write down a general description or even a symbolic generalization for it.

3. Solve the meta-question or describe how to solve it in principle.

Then, perhaps, you can say you understand!

Understanding is not a single all-or-nothing state, but rather a flux, an ebb-and-flow of stronger and weaker connections. Just when you think you understand, something comes along to challenge you further. Without this challenge, you risk becoming stale, staid, and bored.

Learning mathematics is more than doing problems - and far more than merely memorizing or following definitions and proofs. Learning mathematics is a process of using the work of others to guide and inspire your own reconstruction of these ideas for yourself. The pleasures of mathematical thinking come from the energy released when you see how apparently disparate things are in fact related, or when you reconstruct for yourself something that was previously just words or symbols. Thus the purpose of doing exercises is to explore the limits of a mathematical object – seeing how, why, and under what conditions a particular theorem applies or a particular technique works, and when it may or may not be inappropriate.

When you are attuned to these basic processes of mathematical thinking, and to the kinds of questions which mathematics can resolve, you will find your way of looking at the world has been changed for ever.

Training myself (or others) to solve a particular type of question is one thing; educating my awareness of mathematical thinking processes is more complex. Trained behaviour can leave me inflexible and rigid; educated awareness is sensitive to patterns and processes and allows me to think things out for myself. Coupling the two together extends skills which in turn extend sensitivities, which in turn enable further skills to develop in mutual support. That is what Learning and Doing aims to provide.

Learning from Experience, and the Importance of Getting Stuck

Why is the experience gained by doing lots of questions helpful, but not enough?

*One thing we rarely learn from experience
is that we rarely learn from experience alone.*

Learning from doing requires construing - you must make sense of the doing, pausing at intervals to achieve awareness of your actions. This is vital for learning from experience. **The most important thing learned from doing mathematics is not the facts and theories, but the increased sensitivity achieved each time you get stuck**. This means that the next time you get stuck you may think of doing something which this time did not occur to you.

One thing that I have learned from experience is the importance of struggle and the "eureka" moment. To equip myself to 'know what to do' in future I must re-enter as vividly as possible the salient moments when I struggled and eventually found an idea or understood a suggestion that allowed me to proceed. These moments can enrich my awareness and increase the chance that I will think of doing something similar in the future. You can learn from your experience by building on 'things that work', enriching and extending the network of triggers and cues which remind you of strategies that you have used in the past. The foundation of Learning and Doing is gaining experience and then becoming aware of constituent components of that experience.

Preparation for learning

I could say more about doing mathematics, but that would be my experience not yours. More usefully, I can propose tasks which will help you notice features of mathematical thinking which apply to your problem solving, studying, and teaching. It is vital that you

(1) attempt tasks actively

(2) make sense of what you do, and

(3) re-enter and re-experience salient moments after working on each task.

When you have tried something for yourself, you are in a better position to appreciate and make sense of what others have to offer. Therefore, in addition to offering reflective prompts with tasks that I set, I have included my own reflections which I hope make sense to you also because you have worked on that task. Where my reflections speak to your experience, this extends your range of sensitivities and abilities to act in the future.

Mathematical Thinking Processes

Failure can be more useful than success. One challenging problem teaches you far more than many easy problems. Getting stuck gives you an opportunity to learn - when ideas come too readily, you have no marker to return to, no peg from which to extend your network of cues and triggers. The processes of specializing and generalizing - the core of Learning and Doing - are keys to getting unstuck most often. Other useful strategies can be found for example, in *Thinking Mathematically*.[2]

Specializing and generalizing are processes which can be invoked when you are learning and when you are doing mathematics. They are supported by a conjecturing atmosphere, and they contribute to the construction of an argument which convinces first yourself, then a friend, and ultimately a sceptic. Hence you can move from exploration to proof.

In the hands of some teachers, specializing and generalizing have become a mechanical sequence: first you specialize (make a table); then you conjecture a generalization (guess a formula); then you check it on a few examples. This runs completely contrary to the subtlety and interconnectedness of specializing and generalizing, conjecturing and proof advanced in this book.

Every utterance in a conjecturing atmosphere is made in order to bring some thought or idea to expression, to externalize and then consider critically some notion, whether it is an approach to a problem, an insight or proposed argument, or reflections on learning and teaching. I consider this book to be such an utterance, enunciating the conjecture that you will find the ideas here useful not only in your study of mathematics but in developing your view of the world.

John Mason
Woughton-on-the-Green
Autumn 1998

[1] Stein, S. (1996), Gresham's Law: algorithm drives out thought, *Humanistic Mathematics Network*, **13**, p.25.

[2] Mason, J., Burton, L. & Stacey, K. (1984), *Thinking Mathematically*, Addison Wesley, Harlow.

INTRODUCTION

Have you ever found yourself staring at a page of text with no idea what it is about?
Have you ever stared at a blank sheet of paper not knowing where to start?

If so, you are in good company. The aim of this book is to suggest what can be done when you find yourself stuck.

It is not a matter of avoiding what is, after all, unavoidable, but rather developing strategies for doing something about being stuck, when it happens. The idea is to develop *from* the situation rather than be confronted *by* it. Although most of the examples in this book relate to mathematics, the strategies are far more general. The idea is to learn to *do* mathematics, to *be* mathematical, rather than to be confronted *by* mathematics.

The thrust of this book is to focus attention on fundamental processes of thinking. These are not new (you already know how to employ them but you may not always do so when appropriate), nor are they restricted to doing mathematical problems. The same processes are involved in both *doing* and *learning* mathematics, and in learning generally (*mathesis*).

In order to talk meaningfully about mathematical thinking, it is necessary to *do* mathematics. Therefore the text contains numerous questions and problems. They are for *you* to work on – which means getting out pencil and paper, and writing something down.

If you simply read what is written, it will be just words on paper. If you make a serious attempt to investigate each question, then you will have an immediate and vivid experience to relate to the comments. At first you may find that you can write down little more than "STUCK!". But that is fine. By acknowledging being stuck you can free yourself to try to learn what to do next time. The first section introduces something that you can always (well, almost always) do when you find yourself stuck on a problem or a text.

The text focuses on the twin processes of specializing and generalizing, which are fundamental to mathematical thinking, and indeed to thinking of any kind. Specializing means looking at particular cases of a general statement, and generalizing means abstracting features common to several particular examples, but the meanings of both words will expand as the book progresses.

There is nothing mysterious about these processes; indeed, we have all been going through them since birth. Learning to speak, to recognize people and shapes, and to read, are but three examples of "learning and doing".

If the processes of specializing and generalizing are as common as claimed, what is to be gained by drawing attention to them? One answer is that mathematical thinking makes constant use of them, and the more aware we are of what is required, the easier it is to engage in learning effectively. Although they are entirely natural acts, they are more subtle than they first appear. For example, despite their naturalness, we all fail to use them when they would help most. This failure accounts for a lot of the time that we spend being stuck.

Interlaced with the main sections of the book are five short interludes, which explore different ways of talking about studying. The interludes overlap considerably, because I believe that richness and clarity come from a multiplicity of perceptions. The interludes are very much one person's view, though they are based on considerable experience of difficulties that students encounter.

I am extremely grateful to Judith Daniels, who worked through the original version line by line and made many helpful suggestions. Any errors that remain are due to my having inserted them since then! I am also gratified by the many students who have commented that they have found the text useful and engaging. From them has come the impetus to prepare this new edition.

Section 1 SPECIALIZING

The process of specializing is fundamental to mathematical thinking, indeed to thinking of any kind. Basically, specializing means looking at special or particular cases of some general statement, but since it is useful to look at cases which involve only familiar and confidently manipulable objects, such as diagrams, numbers and so on, specializing quickly becomes associated with concrete, confidence-inspiring examples.

The purpose of specializing changes over time. Initially, it is done to try to understand what a statement or question is saying, and later it provides fodder for the reverse process of generalizing.

There is only one sensible thing to do when faced with a claim like the following.

> The sum of the cubes of the first N positive integers is the square of the sum of those integers.

You must try some specific examples, to see what is actually being said.

■■■TRY SOME NOW■■■■■■■■■■■■■■■■■■■■■■■■■■■■■■■■■■

Here is what I found, but this is of no use unless you have already tried the problem yourself!

$N = 1$ The sum of the cube of the first positive integer is 1^3, which is 1.

The square of the sum of the first positive integer is 1^2, which is 1.

Not very inspiring or informative so far!
Now try for $N = 2$ and $N = 3$.

$N = 2$ $(1^3 + 2^3) = 9$.
$(1 + 2)^2 = 9$ seems a bit more interesting.
$N = 3$ $(1^3 + 2^3 + 3^3) = 36$.
$(1 + 2 + 3)^2 = 36$, so at least the content of the claim is beginning to be clearer.

The initial reason for trying these three examples is *not* so much to see if the statement is correct, but to see *what* it is saying; to get it off the page and inside me, so that I can hope to make sense of it. The question of whether it is correct will come later, but in the meantime it must be stressed that just because the claim is correct in those three particular instances, it does *not* mean that it is necessarily always going to be correct for *any* positive integer N.

The entirely natural act of trying specific examples is called SPECIALIZING. It may seem so obvious in this case as to scarcely warrant a name, but there are two good reasons for drawing attention to it. Not only is it rather more subtle in some of its manifestations than may appear at first sight, but also one of the main causes of being stuck while doing a question or while reading a text is *failure to specialize appropriately*.

For the moment, attention should be on the nature of specializing, which in the case above involved using numbers to make sense of a verbal statement.

However, specializing does not always involve numbers. Consider, for example, the formula

$$\cos(A + B) = \cos(A)\cos(B) - \sin(A)\sin(B)$$

There are literally hundreds of formulas like this one in trigonometry, and they can be a bit daunting, to say the least. Yet no mathematician would dream of memorizing them all. For example, specializing by putting $B = A$ gives the formula

| Specializing |

$$\cos(A + A) = \cos(A)\cos(A) - \sin(A)\sin(A)$$

This formula is a special case, or specialization of the first formula, because the scope of the variables A and B has been constrained by the extra condition that A and B are to have the same value. Since this formula can be so easily obtained from the first, it makes no sense to memorize both of them. Similar specializations such as putting $B = 2A$ or $B = 3A$ will yield other formulas, all based on the original. The more adventurous substitution, $A = (X + Y)/2$ and $B = (X - Y)/2$, which is a form of specializing, yields still fancier formulas. It is much easier to remember the main formula, and then specialize to get any of the others when they are needed, than to try to remember them all and run the risk of getting them mixed up.

As a further indication that specializing is not confined to numbers alone, try this next question, and see if your natural reaction is the same as mine.

Tethered Goat

A goat has been tethered with a 6m rope to the outside corner of a 4m by 5m shed. What area of ground can the goat cover?

TRY IT NOW

| Specializing |

My immediate reaction was that I needed a diagram. When I drew it, I found that the question seemed much more concrete. I could see more clearly what was involved, and I found myself extending the drawing to show the various regions that the goat could reach. This in turn showed me what calculations were needed. Drawing a diagram is a form of specializing, in that it is making the question more concrete, more specific, and more manageable.

The *Tethered Goat* example shows that I am using the word 'specializing' to refer to much more than simply trying numbers. It carries a sense of reduction to something simpler, something easier to work with, and so I include under specializing the movement from general to specific, even if the specific is a diagram, or some other representation which makes the question or statement easier to think about. The next example illustrates several sorts of specializing.

In a textbook, the claim is made that the set of numbers (1, 4, 7, 10, 13, 16, ...) does not have unique factorization. It may seem a bit odd to speak of a set as 'having' something, and if it has not been met before, 'unique factorization' may be intimidating. The sensible thing to do is to interpret the technical words in a more familiar context first, such as the set of whole numbers.

Try factorizing some numbers, like 54 or 420, in several different ways and decide what might be meant by 'unique factorization'.

SPECIALIZE NOW

| Specializing |

Working with 54 I found several ways to factor it—to write it as the product of two or more numbers, but when I tried to break it into as many factors as possible, I always ended up with one 2 and three 3's. (I am deliberately not showing my working in order to encourage you to do the work yourself!) This is one sense in which 'factoring of numbers is unique'. Now try the same idea with numbers in the set $(1, 4, 7, 10, \dots)$, but ONLY USING FACTORS WHICH ARE ALSO IN THIS SET.

■■■▶SPECIALIZE NOW◀■■■

I found that 100 is in the set, that $100 = 10 \times 10$, and that 10 is also in the set. But $100 = 4 \times 25$ as well, and both 4 and 25 are in the set. Yet 4, 10 and 25 cannot be further factorized by NUMBERS IN THE SET. So when attention is restricted to this set, unique factorization breaks down.

One point about specializing which is often overlooked, is that it means resorting to particular examples which are confidence-inspiring for *you*. For example, other people's diagrams are never as informative as your own. Thus, what is specializing to one person may be abstract to another—it is all relative to your own experience and confidence. The goat example is pretty straightforward once a diagram is drawn, but it illustrates the general principle that it is often essential to find some way to make a question or a text more concrete. What does it mean to make something concrete? It means to specialize using objects with which you are familiar and which you can confidently manipulate. Sometimes it helps to use physical objects, sometimes numbers are what is needed, and sometimes it may be a matter of reducing complexity by using two-dimensional objects instead of three-dimensional ones, or one-dimensional instead of two. As your mathematical sophistication grows, the 'objects' which give you confidence, and to which you resort for specializing, will become increasingly abstract (as viewed by a novice).

An example of levels of sophistication is given by the following statement:

$$\binom{n}{r} + \binom{n}{r-1} = \binom{n+1}{r}$$

What on earth does it mean? I personally find $\binom{n}{r}$ an off-putting sort of notation, so I immediately specialize by using factorials, which I can confidently manipulate, i.e.

| Specializing |

$$\binom{n}{r} = \frac{n!}{r!(n-r)!}$$

I refer to this as specializing because it turns what is for me a succinct, abstract and not very confidence-inspiring symbol, into something easier to handle. I can now specialize the general formula by interpreting each term of the original statement in the same way:

$$\frac{n!}{r!(n-r)!} + \frac{n!}{(r-1)!(n-r+1)!} = \frac{(n+1)!}{r!(n+1-r)!}$$

It may be that this is no more illuminating for you than the original, but remember that specializing is a relative term. If this is not concrete enough, make it more so! Try writing out what the factorials mean, try specializing further by putting in a specific number for *r*, try specializing even further by giving *n* a value and by confirming that the original expression is indeed satisfied by this choice of *n* and *r*. Effective specializing means using mathematical entities that *You* can confidently manipulate. When I was doing a lot of work with permutations and combinations, familiarity led me to be able to treat $\binom{n}{r}$ as a confidence inspiring object in its own right, just like 37 or $\sqrt{n}$. That concreteness has faded with lack of use.

As a final example of this use of specializing to interpret a statement and make sense of it, consider the following quotation from a textbook.

> Any quadratic graph can be obtained from the graph of $y = x^2$ by means of scalings and translations.

What does it mean? There are some technical terms which may or may not inspire confidence, but the only way to find out is to specialize, to try some specific examples. To do that, it is necessary to sort out the meanings of words, again by means of examples.

What is a quadratic graph? Before concentrating on the word 'graph', it might be best to consider the word 'quadratic':

$y = x^2$ is certainly an example, but it is the one given, so it does not help much!

$y = 2x^2 + 3x - 7$ is a more complicated specific example.

$y = ax^2 + bx + c$ is a general quadratic which, for someone algebraically-confident, can serve to focus attention on the meaning of 'quadratic'. It can be manipulated as any specific 'quadratic' can be manipulated.

The act of writing down some examples has reminded me of my experience with such graphs, so other related ideas are beginning to come to mind—shapes of graphs, graphs being related to each other by geometrical transformations, etc. In particular, the ideas of scaling and translation are reawakened, but if they were not, I would look back and find some examples, some way to write them down in a language with which I feel comfortable

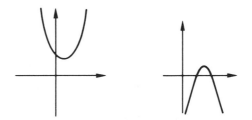

This example reveals another aspect of specializing, which runs throughout mathematics. The quotation focused on the graph of $y = x^2$, not just because it is a simple example of a quadratic, but because in a sense it can stand for, or represent, all quadratics in the same way that 'biro' stands for ballpoint pen. More precisely, every quadratic looks like or can be obtained from $y = x^2$ by suitable scaling and translating. So $y = x^2$ is a very sensible particular example of a quadratic, well worth getting to know intimately. It can then be used as a confidently manipulable example whenever needed.

When the idea of a quadratic is extended to a cubic, it turns out that there is no one cubic that represents all cubics, rather there are three of them. Any cubic looks like, or can be obtained from one of these three, so there are three worth mastering and using as specializations of general statements about cubics.

Whenever someone says 'cubic', these three pictures could pop up on to a mental screen as one of several associations which help to make sense of what is being said about cubics.

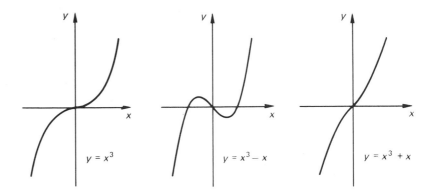

So far, specializing has been recommended as an activity for getting in touch with what a question or statement says. There is a little more to it than that. In order to see what a statement really means, it is necessary to link the particular cases to the general statement which spawned them. In the first example, of the sum of the cubes, the statement does not merely claim that

$$1^3 + 2^3 + 3^3 = (1 + 2 + 3)^2$$

but that, more *generally*, the sum of the cubes of the first N positive integers is always equal to the square of the sum of those integers. The special case gives us a sense of what is meant. Together with other special cases, it suggests a pattern which captures the essence of the original. The understanding comes from the specializing. The process of seeing and describing the general pattern is generalizing.

Summary

It is often tempting to rush into using symbols, only to get bogged down in a morass of vaguely grasped generality, endlessly manipulating the symbols to no effect. When you recognize this state of being STUCK, stop, relax, and accept that you are stuck, and go back and specialize! Knowing that you *can* specialize should begin to provide confidence so that you can treat being stuck as the honourable state that it is, the state from which so much can be learned as long as you do not panic.

Specializing is something which anyone can do in the face of almost any question or text. It is an entirely natural response to meeting an abstract statement, yet all too often it is overlooked. Whenever you find yourself STUCK, ask yourself if you have done enough appropriate specializing.

What does *appropriate* specializing mean? It means specializing drastically enough so that what you obtain makes use of confidence-inspiring, manipulable objects. Sometimes it helps to use physical objects, and sometimes numbers or diagrams are what is needed. As mathematical sophistication grows, so will the complexity of what is considered confidently manipulable:

> numbers,
>> algebraic expressions,
>>> functions,
>>>> set of functions,
>>>>> sets of sets of functions,....

Mathematics is abstract only to the extent that you find yourself dealing with entitites that are not confidence-inspiring—so specialize!

The purpose of specializing is to gain clarity as to the meaning of a question or statement, and then to provide examples which have some general properties in common, so that you can begin to see and appreciate those common properties—the process of generalizing.

Exercises

Work on at least two or three of the following questions, noticing the natural force inside you to specialize.

1.1   Four-square

It is claimed that every number can be written as the sum of at most four squares. Demonstrate this for six consecutive numbers in the thirties. What has your specializing demonstrated about the role of the words 'at most four'? Could four be replaced by three?

1.2   One-sum

Given any two numbers that sum to one, square the larger and add the smaller; now square the smaller and add the larger. Which answer is going to be larger? Make a guess, then try it. Is it always like that?

1.3   What is the square root of 12345678987654321?

STUCK? Specialize to shorter numbers of the same pattern. Use a calculator.

1.4   Tartaglia and De Bouvelles

It is said that in 1556, Nicolo Tartaglia conjectured that the numbers

$$1 + 2 + 4, \quad 1 + 2 + 4 + 8, \quad 1 + 2 + 4 + 8 + 16,\ldots$$

are alternately prime and composite (i.e. not prime).

It is also said that in 1509, Charles De Bouvelles conjectured that one or both of $6N + 1$ and $6N - 1$ (N a positive integer) are prime for all N.

Specialize systematically in order to see what support there is for these conjectures.

1.5   Centre of Gravity

It is claimed that any line through the centre of gravity of a planar two-dimensional object will cut the object into two regions of equal area. Comment.

STUCK? Start with familiar two-dimensional objects. What is it about them that seems to make the claim work? What can be modified to stop it working? Be extreme in your specializing.

1.6 Conversion

I heard someone on the radio say that to convert Centigrade to Fahrenheit in the UK, you need only multiply by 2 and add 30, and you'll be correct to within one degree. Comment.

STUCK? Try some values, and compare them with the true conversion formula $F = 9/5C + 32$. What could be suggested in Sri Lanka, say?

1.7 Divides

If a number divides the product of some numbers, does it necessarily divide one of those numbers? Sort out what is being said, by specializing, and comment on the validity of the statement.

1.8 Unequal

If $x < 3$, then $x^2 < 9$. Specialize, with a view to checking the validity of this statement.

STUCK? Specialize by using numbers. Specialize by using a diagram.

1.9 Annulus

How many equilateral triangles are needed so that when glued together edge to edge they form an annulus, or ring with hole?

example

Interlude A   ON CONJECTURING

Mathematics is commonly thought of as a subject concerned solely with obtaining right/wrong answers. It must be stressed that this is a

MISCONCEPTION!

Certainly, there is an aspect of mathematical work concerned with obtaining particular answers to particular questions, but this is really a very small part of learning mathematics—the tip of the iceberg. All that is visible to most observers is the outer manifestation of answers to questions. The inner activity is like the rest of the iceberg—submerged from view, and by far the largest and most important part.

Take, for example, the notion of a function. It is a technical term in mathematics, and there is a formal definition for it, but when the word 'function' is encountered, all sorts of associations are available to support the formality of the definition. Notions such as

a process typified by the arrow $x \mapsto$,
a graph,
a formula $f(x)$,
some examples,...

all accompany the idea of a function. It is not simply an abstract concept, because it has a component that can be felt in the body as a movement from domain to image, it has a visual component in the form of graphs, and a symbolic component in the form of formulas and the idea of x as a slot to be filled.

There is no right 'sense' of what a function is, though there can be inappropriate or limited senses. Thus for several generations of mathematicians, the sense of 'function' meant a polynomial such as

$$x \mapsto x^3 + 3x^2 - 7$$

It took a long time to accept that something like

$$x \mapsto \begin{cases} x^2, \text{ if } x > 0 \\ x^3, \text{ if } x < 0 \\ 1, \text{ if } x = 0 \end{cases}$$

is also a function. It certainly satisfies the usual definition of a function, since each real number in the domain is mapped to a unique real number in the codomain, yet it can come as a bit of a shock that functions can be pieced together in this way—it opens the door to all sorts of other possibilities! Most of the ideas in mathematics involve multiple ways of looking at something, and there is no one right way.

The fact that mathematics is full of multiple ways of looking at things carries over into the way mathematicians work. Most of the time, mathematicians are very tentative people. They are much more inclined to make a conjecture, that is, to say something which they have every intention of modifying, than they are to make definitive statements that are supposed to be true and correct. This humble attitude rarely comes across in texts, yet many dead-ends may have been encountered and abandoned, and many attempts to say clearly what is meant will have been scrapped before the text is ready. The process of sorting out ideas, whether initiated by a text or by working on a question, involves making conjectures, and being happy to modify them or even throw them out all together if necessary.

Mathematical thinking is best supported by adopting a conjecturing attitude. Never be afraid to offer a tentative conjecture about something, but equally, *do not believe* your conjecture. As soon as I have made a conjecture, I try to write it down or say it out loud. Getting it out of my head and onto the page in front of me helps me to clarify my thoughts. If I try to keep it in my head, it remains woolly and half-formed, and it will certainly clog up my thinking. The struggle to get it down on paper refines and clarifies my conjecture, and clears space so that I can look at it coolly and objectively, checking it on examples to see if it seems reasonable, and trying to find out why it might be inadequate. When it turns out not to be quite right, then I am happy to modify it. That is how mathematicians operate all the time.

The same conjecturing approach applies when working on assessment questions—your rough notes should contain numerous conjectures and modifications. When you have done all you can, it may be that you are uncertain about some parts. The best strategy is to write down your tentative conjecture, clearly marking it as such, and/or to ask your tutor, in writing, for assistance. Tutors marking scripts are always looking for reasons, even excuses for giving marks. They are never trying to find ways of removing them. Put another way, assignments and examinations are opportunities for you to show what you *can* do, not hide what you cannot. If you indicate uncertainties on an assignment, the tutor can be much more helpful to you than if you succeed in covering up your ignorance and fooling the tutor. Thus, in the middle of a computation, a student came across

$$x^2 < 9$$

Uncertain what to do, he wrote

therefore $x < 3$

On paper it looks as though he believes it, whereas in fact it had the status of a weak conjecture. Given that there is not time to think it through by specializing, indicating clearly its status as a conjecture is much more likely to impress the examiner. Furthermore, it is important to gain confidence in your own assessment of what you know. When you feel unsure, it is probably because something is not quite right. The mathematical approach is to register that uncertainty, by labelling the statement as a conjecture. Then if there is time, you can come back and check it out. The chances of fooling the examiner are much smaller than the chances of impressing her!

The purpose of this interlude is to stress that an attitude or an atmosphere of conjecturing frees you from the dreadful fear of being wrong. In fact, it is often better to be wrong so that you can modify your statement and hence your understanding, than it is to be right but perhaps for wrong reasons. In other words, we should bless our mistakes as golden opportunities. Being stuck is an honourable state from which much can be learned. Being right lessens the opportunity to modify and learn. There is always a temptation, when working in a group, to sit back and let others do all the talking in case you make a fool of yourself. In a group which is working mathematically, quite the reverse should be the case. Everyone should be encouraged to express what they have understood, or what they think might be true, so that others can question, and invite or suggest modifications.

Advice

Always mark conjectures *as* conjectures in rough working, assignments and even examinations. That way you will be able to see at a glance what needs further work or sorting out, if time permits, and tutors can be of specific direct assistance. If a group of you are meeting to talk, let it be the group task to encourage those who are *unsure* to be the ones to speak first, and more often than the ones who are sure. Help them and yourself by creating a supportive atmosphere, that is, one in which every utterance is treated as a *modifiable conjecture*!

Section 2 # GENERALIZING

Generalizing is the reverse of specializing, and the two processes go very much hand in hand. Generalizing has to do with noticing patterns and properties common to several situations. A generalization is an expression or statement which can be specialized. It need not actually be true, but it must be able to be specialized to produce the particular cases which it generalizes, amongst others. For example, the sequence of symbols

$$(x, y) \mapsto (-x, y)$$

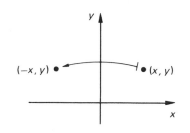

is a way of writing down the effect of a reflection in the y-axis. It captures the essence, the property that is common to all particular instances of a point in the plane being reflected in the y-axis.

Similarly,

$$(x, y) \mapsto (x \cos A - y \sin A, \ x \sin A + y \cos A)$$

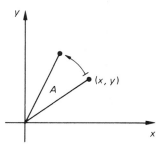

is a general description of the effect of rotating points in the plane, through an angle A anti-clockwise about the origin. It summarizes the effect on any particular point being rotated through any particular angle.

The statement

at $x = -b/2a$, $ax^2 + bX + c$ achieves its maximum value if a is negative and its minimum value if a is positive

is a general statement about the maximum and minimum value of any quadratic expression. It states something which is true about each particular instance, of a quadratic, for example

$$2x^2 - 3x + 4 \qquad \text{has a minimum at } x = \frac{3}{4}$$

$$-3x^2 + 4x - 1 \qquad \text{has a maximum at } x = \frac{2}{3}$$

$$\pi x^2 - x + \sqrt{2\pi} \quad \text{has a minimum at } x = \frac{1}{2\pi}$$

In each case, the sequence of symbols should evoke the awareness of specializations, and also some sort of image.

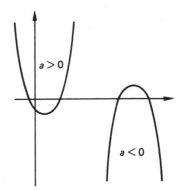

Such general statements make sense, and are of use, only if they act as a catalyst to crystallize experience, fusing together what previously was a lot of disparate examples, both geometric and algebraic. Notice also that a diagram is a particular case, but it illustrates the generality as long as certain features are stressed, such as shape, and certain other features are ignored, such as position and axis-scales.

Often when we feel that we have understood something, it is because we have become aware of connections with other things with which we are already confident. Rather than accumulating a library of facts, it is much easier and more appealing to try to find a general principle which accounts for large numbers of specific instances. This is the role of the most common form of generalizing—the general formula. For example:

Generalizing

- The quadratic formula $x = \dfrac{-b \pm \sqrt{(b^2 - 4ac)}}{2a}$

  gives the roots of the quadratic equation $ax^2 + bx + c = 0$. It is a general statement which can be obtained by generalizing the process of completing the square in any particular instance.

- The expression $2N + 1$
  generalizes the pattern or shape of an odd number.

- The expression $N(N - 1)/2$
  generalizes the computation of the number of ways that a pair of objects can be selected from a particular number (N) of objects.

- The area of a circle $= \pi r^2$
  generalizes results obtained for particular circles by otherwise long and tedious computations involving approximations inside and outside the circle, ideas which led ultimately to the idea of an integral. Because of its succinctness, it stresses the fact that the area depends on the square of the radius.

- The area of a rectangle $=$ length $\times$ width
  generalizes experience with counting squares that tile a rectangle. It applies even when the length and width are not integers but any real numbers.

- Trigonometric formulas like the one for $\cos(A + B)$
  generalize not only specific calculations that can be done with particular angles, but also relationships that hold between angles such as A and $2A$.

- ..., and so on.

Notice that the purpose of this collection is to provide enough data so that you can generalize a common quality: the quality that makes each a special case of a generalization. Whenever the words 'for example' appear, it is a sure sign that a specialization is about to be given, which is supposed to help you to generalize.

The examples given so far suggest that formulas are expressions of general patterns which can be used to recover any of the specific instances which they generalize. Not all generalizations emerge as formulas, however. Techniques for solving whole classes of problems, and the dreaded theorems with proofs which litter mathematics texts, are also statements of general properties common to a class of examples. For example, techniques for solving simultaneous equations, are generalizations of techniques which worked in particular instances and which were then recognized as having wider applicability. When a technique is being learned, it is seen being applied to a few examples, but the general method should be applicable to many other situations. Results such as the Binomial Theorem and Pythagoras' Theorem are also generalizations of observations, synthesizing experience and intuition.

Of course, someone else's generalization is much less likely to have impact or significance than is your own, so whenever you encounter a general statement which seems opaque, or makes you feel uneasy, you should immediately

- specialize;

- try to see what the general statement is saying in the particular instances;

- try to reformulate the statement in your own words as your own generality.

You can usually tell when this process has been successful, because of a good deal of pleasure and lightening of spirits results from capturing a generality for yourself.

The same pleasure can be derived from questions as well as from book-work. Take, for example, the *Tethered Goat* question which was mentioned in the discusion of specializing. It seems like a very dull question, which is easily answered once a diagram is drawn. But what happens if we try setting it in a more general context? We can replace the numbers by letters, but this is only an elementary form of generalizing. More significant questions emerge when we try varying different aspects of the question: imagining ourselves in a field and looking for complications that might arise, and imagining ourselves in an ideal mathematical sort of field and looking for alterations to the conditions. In short, generalizing means placing in a more general context.

■ TRY GENERALIZING IT NOW ■

The idea is to pose questions which involve loosening or changing some of the constraints which are either stated or implied by the original question.

The ideas that occurred to me are listed below. Not all of them are fruitful, or even sensible. There will be time enough to make judgements if and when I decide to try to resolve some of them!

Generalizing

- What about changing the size of the shed and the length of the rope?

- What about obstructions such as trees (mathematical ones would be points, or circles), or other sheds?

- What about a circular shed (I'm thinking of a silo)?

- What about polygonal sheds? (This was suggested to me by the circular shed, because I could use polygonal sheds to approximate a circular one.)

- What about three dimensions? I could look for volumes.

- What about fields with boundary fences as well?

- Instead of asking for the area the goat can reach, I could specify a particular fraction of the whole field of some shape, and ask what length of rope is required. For example, if the goat is tethered to the edge of a circular field, what length of rope is needed to permit it to graze exactly half the field?

These questions will, I hope, give some idea of what is meant by generalizing a question, in the sense of placing it in a more general context.

Finding generalizations which unify previously disparate ideas is the principal source of pleasure in mathematics. It is a bit like finally finding a jigsaw piece which has eluded you for a long time. Putting it in place suddenly brings sense to an otherwise confused picture and a lot of connections suddenly become obvious. Unlike specializing, which is almost always easy to do, generalizing is more of an art, because it involves noticing or stressing things that are common to numerous examples, and ignoring features which seem to be special to only some of them.

One way to experience the effect of stressing, and to create new questions to investigate for yourself, is to take any (mathematical) statement, and to repeat it, preferably out loud, with special emphasis on one particular word. For example, take the statement

Three points determine a circle.

Read it out loud several times, with special emphasis on a different word each time. The effect, because of the stressing, seems to be to raise the question 'why not a different word here?'. Thus

- emphasis on *three* suggests asking what four points would determine, or what special conditions are needed on four points in order to lie on a single circle;

- emphasis on *points* suggests asking if lines could determine a circle, and if so, how many would be needed, or perhaps circles determine a circle (in what way?)...;

- emphasis on *determine* suggests pondering what determine really means, and perhaps modifying or altering that meaning;

- emphasis on *a* suggests moving to several circles, for example, how many points are needed to determine 2, 3, 4,... circles;

● emphasis on *circle* suggests changing to other figures like square, rectangle, and triangle as well as sphere, etc., and asking how many points are needed to determine them, and perhaps even finding some connections between the various answers.

The best way to learn about generalizing is to try it, so I recommend that you take every opportunity, both in this book and in any course you study, to generalize in two ways. First, to replace other people's general statements by your own, as a result of having specialized theirs, so that the generalization is your own. Second, to use the technique of stressing-in-order-to-question, to generate new ideas to investigate.

The stressing technique is overt and calculated to help release implicit constraints brought about by familiarity. There is also an innate force inside everyone to integrate experiences, to try to simplify the complexities of life by categorizing and seeking samenesses—in short, by generalizing.

The following exercise is extremely simple and straightforward, but in many ways it typifies the experience of making a generalization, and shows how deeply ingrained generalizing is.

Dotty

Place a circle around every third dot, starting with the second and continuing from row to row, always reading from left to right.

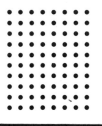

■■■DO IT NOW■■■

I find that I begin by counting out carefully, and after a while something inside me takes over, the counting disappears, and I find myself quickly circling the dots in a diagonal pattern. If you did not notice this happening, try it again on one of the arrays below. Try circling every fifth dot starting from the third, try the same rule but reading right to left on alternative lines, and try your own rule.

The moment of transfer from tedious counting to perceived pattern often happens very quickly. Some people are cautious, and continue counting, steadfastly ignoring or submerging the general pattern, while others, at the other extreme, jump very rapidly to the diagonal pattern, but are prone to making errors because they have not checked their conjecture. The best method is probably somewhere in between, with flexibility.

The simplest generalizations are probably those connected with algebra. For example, the following.

Dinner

An eleventh century manuscript offers the following puzzle.

A lion can devour a sheep in 2 hours
A wolf can devour a sheep in 3 hours.
A dog can devour a sheep in 5 hours.

How long would it take them to devour a single sheep, all eating at the same time? (Ignore the fact that the animals would squabble—medieval puzzles were rarely practical!)

Generalize.

TRY IT NOW

(My resolution is on page 65.)

Doing calculations with particular numbers may have led you automatically to a generalization in which the lion takes L hours, the wolf W and the dog D. If not, try asking yourself what the significance is of the 2, 3 and 5, and whether there is any significance in the fact that $2 + 3 = 5$. Asking yourself about the significance of particulars such as numbers, is another form of stressing-in-order-to-question which can lead to useful generalizations.

Specializing

Generalizing

The authors of the original manuscript would have called it a rule or formula for solving such puzzles. By following the progress of 2, 3 and 5 through the arithmetic without actually *doing* the arithmetical calculations, the general statement can be written down by treating the numbers 2, 3 and 5 as place holders. By using symbols such as L, W and D instead of the numbers, it is much easier to see which places are being held for which things.

More animals can also be added, once the shape of the calculation is perceived. This is what generalization is about. Notice that the answer can be restated as

Generalizing

$$1/\text{TIME} = 1/L + 1/W + 1/D + \dots,$$

which reminds me of the formula for adding parallel resistances in electrical circuits. Perhaps the sheep can be seen as some sort of resistance—an unexpected connection which might be worth pursuing by looking for other questions which give rise to a similar expression. That is what is meant by seeing something in a more general context!

The idea of doing calculations with specific numbers in order to get a sense of what calculations to do in general is not confined to puzzles. For example, the following.

Ratios

If A, B, C, and D are positive real numbers, and if

$$\frac{A}{B} = \frac{C}{D}$$

then

$$\frac{A}{B} = \frac{A+C}{B+D}$$

On first exposure, this sort of question seems quite surprising, and disbelief is best supported (or contradicted), by trying some specific numbers.

■■■ DO SO NOW ■■■■■■■■■■■■■■■■■■■■■■■■■■■■■■■■■

I found that choosing values for A, B, C and D gave me a clue as to how to proceed in general. Taking

Specializing

$\frac{1}{2} = \frac{3}{6}$ gave me $(1+3)/(2+6)$

and from somewhere deep inside me came a sense of the 2 being related to the 1 in the same way as the 6 to the 3, so that adding makes no difference.

I wanted some way to record that relationship, so I decided to let

Generalizing

$$T = A/B = C/D$$

Thus $A = TB$ and $C = TD$, which is how I *felt* the relationship. Having the confidence to try to capture in symbols my sense of what was going on is an essential element at this point. Such confidence can be gained by getting into the habit of specializing so that you always know that there *is* something you can do when you get stuck.

Now I could check that the ratio has the value I wanted:

$$\frac{(A+C)}{(B+D)} = \frac{(TB+TD)}{(B+D)}$$

$$= T$$

This just says, in algebraic language, what I noticed in my specialization.

The only reason for offering this example is that it shows how using numbers can reveal a structure that is not so obvious, unless you already find algebra as easy as arithmetic.

Before dismissing this little question, it is worth testing the extent of your perception of its generality by taking a moment to write down a general statement which includes **Ratios** as a special case. Try the stressing technique. You might also like to investigate what happens if you let the equality change to an inequality.

■■■ GENERALIZE RATIOS NOW ■■■■■■■■■■■■■■■■■■■■

(My resolution is on page 65.)

Investigating the structure of numbers produces all sorts of surprises, and draws upon generalizing in subtle ways. For example:

1, 3, 5, 7,... are all the positive odd numbers;

I can represent any positive odd number by the expression $2N + 1$ (N a positive integer) which stands for a general odd number.

Here, $2N + 1$ generalizes the pattern of odd numbers. The '$+1$' captures the property of odd numbers being one more than an even number, and the '$2N$'

captures the property of an even number as a number divisible by 2. Thus $2N + 1$ is a general odd number. By convention, the N signals the presence or availability of a positive integer (or zero), and in that sense $2N + 1$ describes all positive odd numbers. By convention, N can also mean all integers both positive and negative, and then $2N + 1$ represents *all* odd numbers. If N is read as standing for some *particular* but unspecified integer, then $2N + 1$ represents any particular but unspecified odd number.

Expressions like $2N + 1$ are useful for representing general odd numbers when an argument is sought to show that the sum of any two odd numbers must be even. It is tempting to write $(2N + 1) + (2N + 1)$, but this is not sufficiently general, because it represents the sum of two copies of the same arbitrary but unspecified odd number. The argument requires an expression like $(2N + 1) + (2M + 1)$. The fact that the sum is even emerges because the expression can be rewritten as $2(N + M + 1)$.

There are many other ways of writing down a general description of odd numbers; for example $2N - 1$, $2N + 3$, $2N - 7$ are all equally general, and useful in some circumstances. Now try the following similar example yourself.

Remainders

Write down a general description of all integers which leave a remainder of 1 when divided by 5. Write down a description of all integers that have remainder 3 when divided by 7.

TRY IT NOW

STUCK? Have you specialized first?

Following the model provided by $2N + 1$ for odd numbers, $5N + 1$ and $7N + 3$ are reasonable candidates. The exercise should not be left there, however, because even the simplest of observations can lead to rich mathematical ideas.

More Remainders

Write down a general description of all numbers which are:

- three more than the square of an odd number, and show that they are all divisible by four;
- seven more than the square of a number which is itself one more than a number divisible by four, and show that they are all divisible by eight;
- eleven more than the square of a number which is itself one more than a number divisible by six, and show that they are all divisible by twelve.

TRY IT NOW

The following is an account of my work on these questions, to be read only after you have tried them yourself.

As always I start by specializing:

three more than the square of an odd number

Specializing

$3 + 5^2 = 28 = 4 \times 7$

$3 + 11^2 = 124 = 4 \times 31$

| Generalizing |

$$3 + (2N + 1)^2 = 3 + (4N^2 + 4N + 1)$$
$$= 4N^2 + 4N + 4$$
$$= 4(N^2 + N + 1)$$

The act of specializing on two examples led me automatically to write down a general expression, and simultaneously to the divisibility by four. As with most complex mathematical statements, it is best to start with the component ideas and build up from the inside of the statement, rather than tackling the statement directly from the outside. For example, when attempting to go directly to the general, seek out the embedded term 'odd number', and express it as $2N + 1$; then square it and add three, etc. Applying this inside-out approach to the second statement yields:

$4N + 1$ is one more than divisible by 4

$(4N + 1)^2$ square of...

$7 + (4N + 1)^2$ seven more than...

| Specializing/ Generalizing |

$$= 7 + 16N^2 + 8N + 1$$
$$= 16N^2 + 8N + 8$$
$$= 8(2N^2 + N + 1)$$

To someone experienced and confident with the use of symbols, this *is* specializing. To someone not too confident with symbols, numerical examples may help to reinforce the seeing of a pattern being expressed in the general statement.

But that is not all! The three statements seem plucked from nowhere. Do they have anything in common? What is the same about them?

These sorts of questions are a manifestation of a natural urge to generalize in order to simplify the world. If all three can be subsumed under one common generalization, then they will become more meaningful, more integrated, and more powerful.

More Remainders Still

Express in words and symbols the structure shared by the three statements in More Remainders, and investigate various generalizations.

This example illustrates the richness of trying to see things in more general contexts, in opening them out and seeing what happens if restrictions are removed. It is noticing and articulating patterns which gives generalizing its beauty and its subtlety.

Summary

Generalizing is the twin process of specializing. It involves noticing a pattern common to a number of instances, and trying to express it. It often comes about from trying to see or set a result in a more general context, and this is why generalizing is so intimately connected with understanding. Generalizing is about making connections, and capturing them in a succinct statement from which particular instances can be retrieved by specializing.

Generalizing takes place whenever you stand back from what you are doing to try to distinguish the wood from the trees, trying to place it in a broader context. It is happening whenever you ask for sample questions to be worked by a tutor, in order to try to pick up a procedure for solving similar questions. It is an integral part of understanding, particularly in mathematics, and it is a source of great pleasure.

Whenever you find yourself stuck on a problem or a text, you should first specialize in order to penetrate the technical terms and the wording. Then you should look at the examples you have done and try to see what is common among them, guided by what the problem or text asks for or states. Even if the text is relatively clear, it is much more effective to try to state your own version of it for yourself. There is a great deal of pleasure available by doing this for yourself, and having done it once, it is remarkable how little effort is required to recall it again later.

Strictly speaking, a generalization of some examples is a statement that can be specialized to yield the examples again, but it is useful to carry with generalization the feeling of a 'theme with variations'. The technique of stressing-in-order-to-question is particularly useful. By changing conditions, both explicit ones and implicit ones, and by permitting constants to vary, a statement or question can be placed in a more general context, and interesting questions are likely to emerge. The person who has done the generalizing is the one who will most enjoy pursuing them!

One thing you have to be careful about when you get the generalizing bug, is that you are liable to start asking difficult questions very quickly! Generalizing for the sake of generalizing is like any other drug—you can soon lose touch with reality. The desire to generalize is no excuse for failing to specialize, to become thoroughly familiar with specific examples, since it is only through such intimacy that powerful generalizations will emerge.

Exercises

In each of the following, state a generalization. Make sure that it does, indeed, specialize back to the cases which spawned it. Do not spend time *now* trying to justify your conjectures.

2.1
$$2 + 2 = 2 \times 2$$
$$3 + \tfrac{3}{2} = 3 \times \tfrac{3}{2}$$
$$4 + \tfrac{4}{3} = 4 \times \tfrac{4}{3}$$

$$2^2 \times \tfrac{2}{3} = 2 + \tfrac{2}{3}$$
$$3^2 \times \tfrac{3}{8} = 3 + \tfrac{3}{8}$$
$$4^2 \times \tfrac{4}{15} = 4 + \tfrac{4}{15}$$

2.2
$$7^2 = 49$$
$$67^2 = 4489$$
$$667^2 = 444\,889$$

2.3
$$4^2 = 16$$
$$34^2 = 1156$$
$$334^2 = 111\,556$$

2.4
$$1 + 2 = 3$$
$$4 + 5 + 6 = 7 + 8$$
$$9 + 10 + 11 + 12 = 13 + 14 + 15$$

2.5
$$3^2 + 4^2 = 5^2$$
$$10^2 + 11^2 + 12^2 = 13^2 + 14^2$$
$$21^2 + 22^2 + 23^2 + 24^2 = 25^2 + 26^2 + 27^2$$

2.6 Since $\dfrac{4}{6} = \dfrac{10}{15}$, it follows that $\dfrac{4-6}{4+6} = \dfrac{10-15}{10+15}$.

2.7 Use the stressing technique to extend or generalize the following statements:

- The product of any two odd numbers is odd.

- If the sum of two numbers is fixed, then their product is a maximum when they are equal.

- Given any rectangle, there is another rectangle with the same perimeter but smaller area.

- The sum of any two sides of a triangle is greater than the third.

- The square on the hypotenuse of a right-angled triangle is equal to the sum of the squares on the other two sides.

- The sum of either pair of opposite angles of a quadrilateral inscribed in a circle is the same.

Interlude B # ON CRYSTALLIZING

When an idea or technique is first encountered, it tends to be fuzzy, indistinct and imprecise. Gradually, as further experience is gained, it seems to take shape, until it reaches a reasonably stable form, almost like a crystal. Take, for example, the idea of a number. What does it mean to you?

As soon as I encounter the word number, I find a wealth of associations beginning to flow: some examples; some properties such as the fact that numbers can be added, etc.; and I am reminded that there are different sorts of numbers, such as whole positive and negative, rational and irrational, and so on. The very fact that I have so many associations could be taken as an indication that, for me, a sense of number has crystallized around examples like 2, $\sqrt{2}$ and π. There are also associations with diagrams such as 'the real line'. It is possible that I shall, at some later time, be forced to re-orient my number associations, just as I had to do when I first became aware of the idea of negative numbers, of fractions as numbers, and of irrational numbers. Arithmetic modulo 3, or more generally modulo any integer, also brought about a widening of my number horizons.

How do ideas become associated, and how can crystallization be induced? There are, of course, many facets, but one in particular deserves much more attention than it usually receives. It has to do with activities which I can carry out to help me transform the ideas I meet in a course, into my own ideas. The basic principle is that in the early stages of encountering an idea, while it is still fuzzy, I *see* what it is about, long before I can talk coherently about it to someone else. Even after I can *talk* about it, I find it very hard to *write down* what I understand in a coherent fashion. This interlude is about the transitions from *seeing*, to *saying*. to *recording*.

When generalizing takes place, I find that there is an initial state, which may be very brief or very long, in which I feel that I can see a generalization, but I cannot say it or record it in any way. Subsequently, I realize what I am doing, and I begin to be able to say what the pattern is. It may take several, or many attempts before what I say corresponds to what I think. If challenged to record the pattern in words or symbols, I find it very difficult, though sometimes diagrams make it easier. Even if I write down exactly what I say, it seems to come out very awkwardly. It takes several attempts to arrive at a sensible exposition, and if I try to move to symbols too quickly, I can get myself in a mess.

There seems to be tremendous resistance to making the transition from saying to recording. No matter how much stress a tutor places on getting students to write down notes to themselves about what they are doing when working on a text or a problem, students (and I include researchers) find it hard to do. After considerable observation of myself and others, I have come to the conclusion that there is a great difference between spoken speech and written records. Speech is halting and disjoint in the sense of mostly unfinished sentences, and it is constantly monitored and modified. Writing, on the other hand, demands complete sentences, and does not respond well to frequent qualifications and modifications, despite what one might expect. In other words, our reluctance to write things down is not simply due to laziness or some other obduracy, but reflects an inherent difficulty in switching modes.

It seems reasonable, therefore, to pay some attention to the transitions from seeing something, to saying what I see, to recording it. In particular, it often helps in the latter transition to use pictures and bits of sentences before trying to get complete thoughts, and especially before trying to put everything into symbols. It will be much easier to make records if I am articulate about what I have seen, so it is well worth spending time telling and re-telling someone (or something, like a goldfish or an imaginary friend walking with me to the bus stop!) how I see what is going on. The attempt to talk about some idea, and then to write it down, is an attempt to capture or crystallize it in some form so that it gets out of my brain. Then I can look at it afresh. So often, ideas can lie in a pre-articulate woolly phase, and then when I need them, I find that it all slips away. (How many times have you thought you understood something only to find a day or so later that it has all escaped?) Research mathematicians know that the very act of trying to tell someone else your problem helps to clarify things, and often the listener need say nothing.

Advice

Whenever you encounter an idea in a text, try re-stating the idea in your own words. At first, you may find it hard to use words other than the ones in the text—if so, specialize to some examples, and tell yourself, out loud if at all possible, in what way the specializations are special cases of the more general idea. At the end of each section, try to tell yourself what you think the section has been about. When you have had several goes at saying it, try to capture it as succinctly as possible, using symbols, pictures and a few words. Refer back to these every so often, as you need them. Why not try it now with this interlude, and with the previous section?

Section 3   Specializing and Generalizing Together

Although specializing and generalizing were introduced separately, they are often hard to keep separate. The reason for specializing is to permit and to promote generalizing. Generalizations need to be checked in specific instances before looking for a convincing argument. To show how intimately they are bound together, this section begins by investigating the famous theorem of Pythagoras, which has been the starting point for many different strands of mathematics.

Pythagoras' Theorem

In any right-angled triangle, the square on the hypotenuse is equal to the sum of the squares on the other two sides.

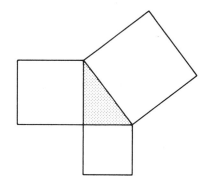

Symbolically, this can be written out as the following.

If *C* is the hypotenuse of a right-angled triangle, and *A* and *B* are the other two sides, then

$$A^2 + B^2 = C^2$$

There are several differences between the two versions. To begin with, the first version speaks of 'the square on...' meaning 'the area of a square with side...', but this is entirely lost in the second version. The second uses symbols, and most of the statement is taken up by naming the symbols and indicating their role. Even so, this version fails to state that *C* is actually the length of the hypotenuse, and similarly for *A* and *B*. Most people remember the equation as the theorem and then reconstruct what the notation means, if necessary.

Even though the theorem bears the name of Pythagoras, who lived in the sixth century BC, there is considerable evidence that it was known, at least in some form, at least a thousand years earlier by the Babylonians, who stored tablets in their library containing at least fifteen specific numerical cases of the theorem, such as

$$3^2 + 4^2 = 5^2$$
$$5^2 + 12^2 = 13^2$$
$$8^2 + 15^2 = 17^2$$

We do not know whether there were any pictures or diagrams to accompany the table. The fragment that remains contains integer solutions to the Pythagorean equation, and they are sufficiently extensive and organized to suggest that the Babylonians had some procedure for generating numerical examples, even if they did not have an explicit statement of the geometrical theorem. We also know of a Chinese geometrical instance of the theorem, of uncertain date, but at least as early as Pythagoras.

Notice that there is a big jump from having a table of numerous instances to a statement which covers all real numbers, or to a geometric proof. It is a matter of debate among historians as to whether, in our terms, the Babylonians had a generalization, and if so whether it was geometric or algebraic or both, or whether they simply had over the years, found many specific instances.

Pythagoras' Theorem is by no means the end of the story as far as generalization is concerned. Each of the following statements suggests a more general context for looking at the theorem, derived from stressing individual words in the statement of the theorem. Some of them suggest a variation on a theme, some suggest a generalization of the theorem, and some are generalizations as stated. You might like to decide which before reading my comments.

A In any triangle, $c^2 = a^2 + b^2 - 2ab \cos(C)$, where a, b, c are the lengths of the sides and C is the angle opposite side c.

B If a regular pentagon is drawn on each edge of a right-angled triangle, then the area of the pentagon on the hypotenuse is the sum of the areas of the other two.

C $3^3 + 4^3 + 5^3 = 6^3$

D The square of the length of the diameter of a cuboid with sides a, b, c is $a^2 + b^2 + c^2$.

E In any triangle, $c^2 = a^2 + b^2 - ab \cos(C)$, where a, b, c are the lengths of the sides and C is the angle opposite side c.

F $2^2 + 6^2 + 9^2 = 11^2$

G If x, y and z are positive integers, then $x^n + y^n = z^n$ implies that n must be 2.

H If, in a tetrahedron, there is a vertex with three right-angles adjacent to it, then the square of the area of the face not containing a right-angle is equal to the sum of the squares of the areas of the other three faces.

I Place an even number of points uniformly round a circle of unit radius. For any point P on the circle, the sum of the squares of the distances from P to each point is independent of where on the circle P is chosen.

A generalizes the idea of a right-angled triangle to any triangle, thus releasing the restriction to right-angles in the original statement. Specializing **A** by choosing angle *C* to be 90° gives the Pythagorean equation as a special case. This is typical of a generalization.

B changes the context by replacing the geometric image of squares being constructed on the edges of a right triangle, with pentagons. If may seem implausible at first sight, in which case, try replacing pentagons with equilateral triangles and doing the calculations. Experience with squares and equilateral triangles suggests the further generalization that as long as the figures are similar (that is, scaled versions of the same shape), then the areas add as suggested. This is close in spirit to the way Euclid and his friends are believed to have perceived Pythagoras' Theorem.

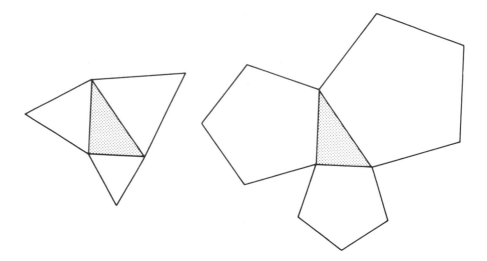

C suggests generalizing the idea of adding together squares to adding together cubes, and provides one instance. Adding cubes to get a cube is a variation of Pythagoras' equation rather than a generalization, because it does not specialize back to the original. Neither does it indicate any analogue for the geometrical part of the theorem. Mathematicians have discovered a wealth of interesting questions by following up the numerical aspect while ignoring the geometric, and vice versa.

D generalizes the two-dimensional aspect of Pythagoras to three dimensions. Specializing one of *a*, *b*, *c* to be 0 reduces **D** back to the two-dimensional triangular case.

E generalizes the right-angled triangle notion to other triangles. Specializing by taking angle *C* to be 90° gives the Pythagorean theorem. *However*, this generalization, while being a reasonable first conjecture, is *not* valid for all triangles, as it stands. For example, it fails for any equilateral triangle (specializing!). It is a generalization, but not a true assertion. Unlike statement **A**, it fails to give correct answers for other values of angle *C*.

F suggests a generalization from the sum of two squares to the sum of three squares, but offers no geometric side to the analogy. It specializes back to the Pythagorean equation by making one of the squares zero.

G generalizes the idea of integer solutions to the Pythagorean equation to sums of higher powers. This is the famous *Fermat* conjecture, which has so far resisted all attempts to provide a convincing argument that it is true. Who knows, it may be false!

H generalizes the whole Pythagorean theorem to three dimensions, and suggests extending it to higher dimensions still!

I suggests a more general context which glues together several instances of the original theorem (since a pair of diametrically-opposite points forms with *P* a right-angled triangle). This, in turn, suggests investigating further generalizations—why not an odd number of points; must the points be uniformly spread around the circle; and what happens in three dimensions?

Generalizations lead to conjectures, which may turn out to be true, or false. The first thing to do is to check that they do, indeed, specialize back to your particular cases. Conjectures can then be investigated further by specializing in order to see not *what* might be true, but *why* the conjecture might be true, or *why* it might be false and so need modifying. Suggestions for trying to build convincing arguments to support conjectures are given in the next section.

There are many other general questions which stem from the same Pythagorean Theorem. For example, in order to generate other specializations of

$$A^2 + B^2 = C^2$$

with *A*, *B*, *C* all integers, it would be tedious to have to explore by trial and error. In fact, it is possible to state a generalization of the three examples

$$3^2 + \ 4^2 = \ 5^2$$
$$5^2 + 12^2 = 13^2$$
$$8^2 + 15^2 = 17^2$$

in the form

$$(M^2 - N^2)^2 + (2MN)^2 = (M^2 + N^2)^2 \text{ where } N \text{ and } M \text{ are any integers.}$$

By specializing this statement, we can recover the particular instances.

TRY $M = 2$, $N = 1$ and $M = 3$, $N = 2$ YOURSELF, NOW!

What values of *M* and *N* will recover the 8, 15, 17 instance?

(My resolution is on page 70.)

Specializing has demonstrated that the general statement includes the particular instances, but it tells us nothing about whether the generalization is always valid. In this case, it only requires some algebra to demonstrate that

| Generalizing |

$$(M^2 - N^2)^2 + (2MN)^2 = (M^2 + N^2)^2 \text{ where } N \text{ and } M \text{ are any integers.}$$

Expanding each of the terms on the left-hand side we have

$$M^4 - 2M^2N^2 + N^4 + 4M^2N^2 = M^4 + 2M^2N^2 + N^4$$
$$= (M^2 + N^2)^2$$

Given that $M^2 - N^2$, $2MN$ and $M^2 + N^2$ provide examples of what are known as Pythagorean triples, i.e. triples of integers which satisfy $X^2 + Y^2 = Z^2$, there are at least three questions which arise by seeking a more general context.

- Can *all* Pythagorean triples be generated by the expressions in *M* and *N*?
- Can *any* number appear in a Pythagorean triple?
- Can similar expressions be found for generalizations such as the sum of three squares being a square?

The first provides a useful example of what can go wrong when you are trying to produce a convincing argument, so I shall defer considering it until Section 5. The third appears briefly in the exercises at the end of this section. I shall investigate the second. Do not worry unduly about the technical details. Rather, pay attention to the uses of specializing and generalizing, and the dead-ends encountered along the way, for they typify what it is like to *do* mathematics as distinct from studying polished mathematics texts.

Can *any* number appear as one of $M^2 - N^2, 2MN$ or $M^2 + N^2$? I immediately looked at $M^2 - N^2$, and said to myself that surely any number can be written as the difference of two squares. A tiny bit of specializing showed that my conjecture would need modifying.

	Specializing		

M	N	$M^2 - N^2$
2	1	3
3	1	8
3	2	5
4	1	15
4	2	12
4	3	7
5	1	24
5	2	21
5	3	16
5	4	9

I do not see how 2 can arise as the difference of two squares. I have obtained 3, 5, 7 and 9, which suggests that I might (being more cautious now!) be able to obtain any odd number. Before diving in further, I pause to recollect what I WANT. I want to know whether *any* number can appear in some Pythagorean triple. How else might I proceed?

Why not look at the $2MN$ term instead! Given any number, I can factor it, and surely I can choose M and N as two factors. Try 6 for example. I know that $6 = 2 \times 3$, so I can take $M = 2$ and $N = 3$. CHECK! No, I need to have another 2. Try $M = 3$ and $N = 1$. Then the triple will be

$$3^2 - 1^2, \quad 2 \times 3 \times 1, \quad 3^2 + 1^2$$

which is

8, 6, 10

The 2 in $2MN$ is a bit awkward because it produces an even number. What shall I do if I am given an odd number to put in a triple? I shall have to resort to the difference of two squares, but I feel hopeful about odd numbers as differences of squares.

Specializing

I WANT to express any odd number, such as $2K + 1$, as the difference of two squares. Specializing, using some of the earlier data:

$$1 = 1^2 - 0^2$$
$$3 = 2^2 - 1^2$$
$$5 = 3^2 - 2^2$$
$$7 = 4^2 - 3^2$$

A pattern is beginning to emerge. Generalizing from this systematic specializing, I try

Generalizing

$$2K + 1 = (K + 1)^2 - K^2$$

Immediately, I recognize that this is right because I am familiar with the binomial theorem, and I realize that I might have seen it earlier if I had paused for a moment. I have not quite finished however, because I want to exhibit M, N and the Pythagorean triple.

$$M = K + 1 \quad \text{and} \quad N = K$$

and the triple is

$$2K + 1, \quad 2K(K + 1), \quad 2K^2 + 2K + 1$$

Have I answered my original question? I have implicitly, but not explicitly. Given any positive integer (I am being careful now), that number can appear as a member of a Pythagorean triple: if the given integer P is even, then M and N can be found by factorizing P as $2NM$, and if P is odd, then M and N can be chosen as $K + 1$ and K where $P = 2K + 1$ as shown above.

It is always a good idea to pause when you reach a conclusion, and look back over your work. The details of the argument need to be checked, because it is terribly easy to make an error in a calculation, or to overlook some possibility. It is also important to see what can be learned—for example, in this case I responded too quickly to the first idea that came into my head, and narrowly saved myself from a long diversion considering which numbers can be written as the difference of two squares. It is a worthwhile question to pursue in its own right, but it is not necessary for the original question, as in the end I needed only the special case of representing an odd number as the difference of two squares, which I was able to resolve quite easily. It also occurs to me to investigate which numbers can be expressed as the sum of two squares, and I will certainly need to know that in order to tackle the following.

In how many different Pythagorean triples can a given number appear?

The point is that when one question is resolved, there are usually plenty more arising from it which help to put the original in a broader context. Questions like 'How many different...?' and 'What is the maximum/minimum...?' are typical mathematical questions.

It is important to realize that, despite the neat and orderly appearance of textbooks, *doing* mathematics entails making plenty of false conjectures, and pursuing many dead-ends. Textbooks omit them because it is not very often instructive to follow other people's dead-ends, indeed it can be intensely frustrating, and surely the purpose of a text is to show how something *can* be done, not how it cannot. In your own studying, and in tackling questions, you will encounter many false trails—do not be disheartened, but rather take pleasure in the hunt, and in the gradual refinement of conjectures. When you reach a statement, specialize, preferably systematically, keeping all the while an eye on the generalization you are after.

Here is another example of a question which invites several different approaches, though most of them are rather hard to follow through!

All Ones

Which of the numbers 1, 11, 111, 1111, 11111,... can be a perfect square?

■■■TRY IT NOW■■■■■■

STUCK? Try them on your calculator! Make a conjecture. Which digits do perfect squares end in?

Do not spend a long time on this question! It is not worth a lot of effort, apart from seeing the many possible dead or at least difficult 'ends'.

Resolution

Trying the numbers on a calculator seems a good place to start. The calculator shows that only 1 seems likely to be a perfect square, but it is far from clear why this might be.

The terms 11, 1111, 111 111,... are divisible by 11, which suggests looking at the numbers 1, 101, 10 101,... obtained by removing the factor of 11. These numbers would have to have a factor of 11 as well, if the original number was to be a perfect square—a lot of energy can go this way!

The square roots on the calculator are

1, 3.32, 10.54, 33.33, 105.41, 333.33,....

I could look for a pattern here, trying to make use of the appearance of those threes.

If I stop and ask myself which digits can be the last digits of a perfect square, I find that numbers ending in

1, 2, 3, 4, 5, 6, 7, 8, 9, 0

when squared, have as their last digits

1, 4, 9, 6, 5, 6, 9, 4, 1, 0

Only numbers ending in 9 or 1 could possibly be square roots of numbers consisting solely of 1s. Dead halt—what now?

Carry the same idea further—what two-digit numbers can be the last two digits of a perfect square? Careful—that looks like a lot of work. I really want to know whether 11 can be the last two digits of a square. Trying all the cases as I did for single digits *will* get an answer, but so will a little algebra!

Suppose $10A + 9$ or $10A + 1$ were square roots of an all 1s number. Their squares are

$$100A^2 + 180A + 81 \quad \text{and} \quad 100A^2 + 20A + 1$$

I want the second-last digit to be a 1, but the second-last digit comes from the last digit of $18A + 8$ or from $2A$, which in both cases is even! None of the all 1 numbers can be squares.

Checking over the argument, I wonder where the fact that 1 *is* a perfect square shows up?

Generalizing

I have also shown a great deal more than I wanted originally, because the argument looks at only the last two digits. It never occurred to me at the beginning that only the last two digits mattered—perhaps I was blinded by the plethora of 1s.

The point of the examples is not that they are yet more mathematical content to be learned, but that they illustrate a number of points about specializing and generalizing.

- Specializing and generalizing go hand in hand. They cannot easily be separated.
- The purpose of specializing is first to give substance and meaning to abstract statements, and then to find out why something might be true, or why it is false. It can be shown to be false by finding a counter-example, and then more specializing is needed to see how to modify the conjecture.
- Calculations with specific numbers often suggest what calculations to do with more general symbols.
- Setting a result in a more general context can lead to interesting questions, and supports understanding by showing up previously unsuspected connections.

The Pythagorean example also demonstrates how a mathematical investigation might be carried out. Whereas most mathematical texts consist of well worked-over and tidied solutions which frequently have lost all the flavour of the original discoveries, the account given above was just as I worked it out, without editing. Looking back over it you will notice the frequent use of I KNOW and I WANT, which I find useful in order to keep track of where I am and where I am going. Their use will be developed further in later sections.

The next short example illustrates how specializing and generalizing can help each other, by showing that it often helps to do both at the same time when working on a question.

> If P is any point inside a regular N-sided polygon, then the sum of the distances from P to each of the edges is independent of where inside the polygon P is placed.

The first thing to do is to specialize.

| Specializing |

■■■SPECIALIZE NOW■■■■■■■■■■■■■■■■■■■■■■■■■■■■■■■■■■■■■

I am not going to show you all of my diagrams, because you must draw your own—yours can speak to you whereas mine have to be interpreted and worked on.

Diagrams are a very useful form of specializing, or perhaps more accurately, of particularizing. For example, imagine a square. Now draw a square on a piece of paper. The square in your head may be very clear, or rather fuzzy and unstable, but it probably does not have an actual size. The square you have drawn has a size, but is probably not exactly a square. Yet by looking at your diagram, you can stabilize your mental image of a square, and look through or past the particularities of your drawn square, as if you were seeing all squares, or perhaps 'squareness', rather than a single specific square. An actual diagram is particular, but it can be looked at as if it were generic—as if it represents the general case. Thus attention focussed on the diagram can look through the particular to the general. When specializing, it is sensible to look at cases in which the calculations are reasonably easy, like $N = 4$, $N = 6$, and indeed, as I

| Generalizing |

conjectured and then checked, $N =$ any even number (note the generalizing). Finding a justification for the full claim involves being able to do the calculations in the awkward cases as well, like $N = 3$ and $N = 5$. I twice started some calculations but gave up almost immediately because they seemed so unpleasant.

Before launching in, it is worthwhile considering what is wanted and what is known. In this case, it is known that the polygon must be regular—a rather unsatisfactory state of affairs because I immediately get curious as to why that condition is put in. I tried 3 points (the simplest case), but without the regularity. I couldn't see much, so I tried 4 points without regularity, and it became clear to me that the claim works for the regular case because you can easily do the computation, but as soon as you have irregular figures, the calculations are going to get caught up with the lengths of the sides of the polygon.

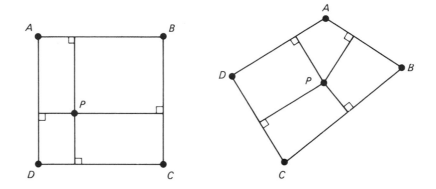

That led me to ask how the distances of P to the sides, and the lengths of the sides, might be connected...and suddenly it popped into my head that their product gives the area of a triangle...well, double the area. More specifically (as I looked for a way to record what I had seen), if A and B are two adjacent vertices of the polygon, then the perpendicular from P to AB, times the length AB, is twice the area of the triangle PAB.

In a flash, I saw the original polygon divided up into triangles with P as a vertex common to them all. If the edges of the polygon are all the same length, then the sum of the areas is the sum of the distances from P to the edges, times the constant edge length. Now I could see what the question was getting at.

I can begin to see how I must be a bit careful about whether the polyon is allowed to cross itself, since then areas might overlap, and I have a vague taste of a possible generalization which permits P to be anywhere, not just inside the polygon.

Again, the purpose of this example is to illustrate features of specializing and generalizing, not to carry the investigation to completion. In this case, the act of generalizing at the same time as specializing led me to see an argument that would avoid fearsome computations.

Specializing and generalizing have, so far, been illustrated as processes for coming to grips with questions or with texts. They actually play an even more fundamental role, for where does the content of mathematics texts come from? The techniques, definitions and results all come as the result of people asking questions, and then trying to answer those questions, however partially. Studying mathematics is not just a matter of studying other people's solutions to other people's questions. Mathematics as a subject comes alive for you when you start noticing your own questions. There are questions around all the time—and the ability to notice them is tied up with seeking generalities. It all has to do with looking at a *particular* situation, and seeing it as representative of a generality. Most questions have the quality of *why* or *is-it-always-so*, which are based on generalizing from the particular.

Whenever you encounter a mathematical or physical situation, asking the questions

- What if...is changed?

- What happens if...?

- Of what is this a special case?

- In how many different ways...?

- What is the most/least...?

may lead to mathematically rich investigations. 'What if' type questions are often stimulated by the stressing technique. They are equally applicable when trying to understand a passage in a text, working on an assignment and even when walking down the street. Used in a text, they can direct attention to other ideas and thereby bring out connections, helping you to place what you are reading in a broader context. Used with an assignment, they may suggest relevant ideas that help you with the problem at hand. Used 'out of doors' or 'in the market place' they can lead to fascinating and difficult questions.

For example, the following are just a few that have happened to me.

Do not spend time investigating these now. A better use of time would be to start collecting your own!

Night-rider

Riding on a cycle-path one night, I overtook another cyclist, and then went under a street lamp. I was surprised by the shadow of a head, racing past me, so I swerved to avoid what I thought was the other cyclist trying to pass me. I glanced behind me, only to find him well behind. What was going on?

Tanked

Walking near a large oil storage tank, I noticed a set of stairs attached to the side of the cylindrical tank and rising from bottom to top. One end of each step was attached to the tank wall, and other end was free, so I had a fine view of the shape the steps made. I was surprised because the shape was not what I expected.

Double Glazing

Watching the sun rise one fine morning, I noticed that I could see two suns—a very bright one and a weak one. The weak one appeared just below the bright one, and as I rocked from side to side, the weak one seemed to swing from side to side relative to the bright one. The window was double-glazed, the panes being some 3 inches apart. On another occasion, at a different window, the weak sun appeared to move up and down. Explain!

Umbrella

Walking in a real downpour with an umbrella, I managed to keep my head dry, but little else. How large should an umbrella be in order to be effective, assuming that it is used sensibly?

Freezer

I have noticed that it is very difficult to open the freezer door immediately after having closed it. If I wait a short while, it is much easier. Explain!

See-saw

Walking in a playground, I noticed that the see-saws were not like the ones from my childhood, because they remain horizontal when no one is on them. What is the mechanism, and what is the path of the seat as it goes up and down?

It is important to remember that most of the questions generated in this way will be hard—they need to be specialized by simplifying and experimenting, and in most cases the question needs clarifying before it can be worked on. The point is that by *noticing* questions, you exercise your powers of specializing and generalizing.

As a final example of the interplay between specializing and generalizing, consider the following question.

Constant Perimeter

For which rectangular boxes do all faces have the same perimeter?

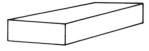

■■■TRY IT NOW■■■■■■■■■■■■■■■■■■■■■■■■■■■■■■■■■■■■■

STUCK? Specialize—make sure you know what the terms mean, then look for an example. Now look for another. Can you detect anything going on? Try to express it in general terms so that it applies to all rectangular boxes.

| Specializing |

I found 'cube' pretty quickly, but each time I tried to produce another example, I couldn't get the third face to have the same perimeter as the other two. (There are six faces, but they are identical in pairs, so really there are only three I need think about.) For example, if the first face is 2 by 3, then it has perimeter 10. The second face has to be 3 by something with the same perimeter, so it must be 3 by 2, but then the third face is 2 by 2 and has the wrong perimeter.

Trying to express in words my sense of what is going on, it came out rather clumsily. Once you have chosen a side and a perimeter, the second side of two faces is determined, and must be equal, so the third face is a square. It cannot have the same perimeter as the other faces unless they are all square. I find that hard to follow, and it's my own argument!

Symbols are much clearer. (My symbols are much clearer to me!) Suppose the faces of the box are P by Q, Q by R, and R by P. The perimeters are $2(P + Q)$, $2(Q + R)$, and $2(R + P)$, which I know must all be equal. Equating the first two gives

$$2(P + Q) = 2(Q + R)$$

so

$$P = R$$

and equating the second two gives $Q = P$, so the box must be a cube.

Notice how the succinctness of the algebra communicates so much more clearly than the words—even if you refine the words and say it better. The algebra actually helps the pattern or structure to emerge. As usual, **Constant Perimeter** cannot be permitted to finish there. The stressing technique suggested to me to try other shapes. I immediately thought of a tetrahedron.

■■ TRY IT NOW ■■

(My resolution is on page 70.)

One nice thing to emerge from the resolution of the tetrahedron variation is that it leads to yet another conjecture: given any triangle, then a tetrahedron can be constructed with four congruent copies of that triangle as the faces. I invite you to pursue this in the exercises. Specializing with scissors and paper will suggest that this conjecture needs modifying. When a new conjecture arises which seems very plausible, if not downright convincing, how can someone else be convinced? That is the subject of the next section!

Summary

Specializing and generalizing are intimately tied together. Specializing can produce fodder for generalization, and generalizations must be checked to see that they do specialize back to the particular cases which spawned them. Both specializing and generalizing are more subtle than they first appear. Often, it is important to specialize systematically so that a pattern can emerge. At other times it is helpful to be extreme, to stretch an idea to its limits, in order to see what is going on. The overall purpose is to become aware of, and to enunciate conjectures which can then be examined and either modified or justified—the topic of the next section.

Generalizing also involves reflecting on a section of a text or a question, trying to place it in a broader perspective, to see links with other ideas or questions. By setting yourself to look at particular events or situations and see them as indicating or suggesting a more general phenomenon, you can begin to ask your own mathematical questions. Mathematical thinking may then become alive inside you, and all this 'talk' about specializing and generalizing will become second nature.

Exercises

3.1   It has been claimed that if you have four congruent triangles, then these can form the four faces of a tetrahedron. INVESTIGATE!

STUCK? Specialize to concrete materials—scissors and card or paper. Specialize your triangle to be equilateral, then vary it. What sorts of things might go wrong? Try extreme special cases.

3.2   Place the following arithmetic facts in a more general context,

$$1^2 + 2^2 + 2^2 = 3^2, \qquad 4^2 + 4^2 + 7^2 = 9^2$$
$$2^2 + 6^2 + 9^2 = 11^2, \qquad 4^2 + 6^2 + 12^2 = 14^2$$

by verifying that for any numbers X, Y and Z,

$$(X^2 - Y^2 - Z^2) + (2XY)^2 + (2XZ)^2 = (X^2 + Y^2 + Z^2)^2$$

and specializing. Show that $2^2 + 3^2 + 6^2 = 7^2$, which is the fourth example given with a common factor removed, and that $3^2 + 4^2 + 12^2 = 13^2$, but that *no* choice of integers for X, Y and Z will yield these particular cases.

3.3   How many dots will be needed to make the 37th diagram in the following sequences? Generalize.

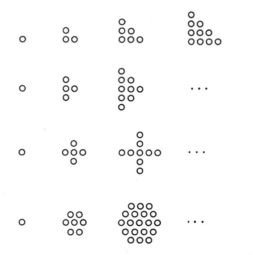

STUCK? Specialize by counting the dots, then look for a pattern, then see if the dots can be rearranged to demonstrate your conjecture. Make up your own patterns.

3.4   Family Tree

I heard someone on television claim that he was one-third Cherokee. He is in good company, for in the oldest known recorded story, the Sumerian tale of Gilgamesh, it is claimed that Gilgamesh is two-thirds god and one-third man. Investigate these claims.

STUCK? Specialize, perhaps using a family tree. You WANT to find out how such calculations are done, and then how one-third can be an answer.

3.5   Crossless

In the plane, P points are to be joined in pairs by straight-line segments, but they may not cross each other. What is the maximum number of line segments there can be?

Specialize: draw your own diagram.

STUCK? Reformulate a more precise question for yourself. Specialize. Do not believe your conjectures, explicit or implicit!

Interlude C   ## ON EXPECTATION

Studying mathematics can be very frustrating. Some new ideas are encountered, with techniques for solving certain types of problems; there is barely time to get a sense of the ideas before it is time to practise the techniques on some exercises, and then do some assignment questions. Often, it is possible to use the examples and exercises to master the techniques and score good marks on the assignment, yet a few weeks later the ideas and the techniques seem to have evaporated. Even if the techniques are recalled, there may be little satisfaction or sense of understanding. In other words, mastery and understanding may not be quite the same things. I can think I understand something, yet not be able to do the examples; alternatively, I can be able to do exercises perfectly, and not understand much about what I am doing. It is worthwhile, therefore, to clarify expectations about understanding and mastery.

There is one simple lesson that many people find difficult to accept:

I CANNOT expect to MASTER everything on first exposure.

Indeed, it is not often possible to master *anything* on first exposure. It is a bit like being on board ship, pulling into harbour in a fog. A distant foghorn is heard intermittently, but nothing can be seen through the fog. Gradually, the foghorn gets louder, and vague shapes appear in the mist. Finally, the ship pulls up to the dock, and I can get onto firm ground.

The initial foghorn is like initial exposure to ideas—it gives me a sense of direction, a vague notion of what is happening. As I gain experience working on examples, vague shapes emerge from the mist, and I begin to see what is going on. With continued exposure, I begin to discern more details, to see more of what is involved. Eventually, I become aware that it is time to practise the technique or to use the idea in new contexts. With sufficient practice, the techniques and ideas act as firm ground for the growth of understanding. The most important aspect to remember is that it does take time, and that understanding will grow.

In the case of a particular technique, on first exposure, I try to find out WHAT the technique does. With continued exposure to worked examples, I begin to see HOW it works, and begin to get a sense of WHY it might work. In the pressure to keep up with my studying, it is easy to forget that the worked examples are specializations of something, and that the text gives me guidance to the generalization. Thus, any examples I work on should be related directly to the general statements in the text, and not just worked through on their own. Eventually, I reach a point where I recognize that I must get down and master the technique by doing lots of exercises myself. Trying to do this too early is often a waste of time, and putting it off just clogs up my memory, because once I have really concentrated on mastering the technique, I no longer have to remember it. It becomes second nature.

Sometimes, I find myself short of time, so I use the worked examples and exercises as indicators of how to perform a technique. This is a perfectly mathematical approach, since it is using specializations provided by the text to pick up clues about the general technique. However, too much of this is a waste of my time and mathematical skills, because I am ignoring the text which actually gives guidance as to the generalization. Furthermore, it is all too easy to focus on one salient aspect and miss what the technique is really all about. The result is more often frustration and uncertainty than pleasure and confidence. In the short term, it may get an assignment done, but it does not always lead to understanding.

The pressure of new work is always present when studying mathematics, and it is compounded by the pressure, mounting to hopelessness and panic, of previous work only partly comprehended and insecure. The best way to reduce pressure is to adopt a sensible level of expectation. The pace and type of activity may seem to be dictated by the course materials, and to some extent this is true, but the feeling of being driven forward at breakneck speed is largely due to expectations. If I expect to master everything on first encounter, then I will definitely feel great pressure. If, however, I bear in mind the necessity of a gradual development from first seeing, through increasing exposure, to final mastery, I can relieve a good deal of the pressure. That does not mean that I can sit back and expect the ideas to mature miraculously inside me—I need to keep struggling with the ideas—but being aware of the need for time and repeated exposure can make studying easier, and much more pleasant.

Mathematics is often portrayed as a linear subject that develops relentlessly, so that I cannot understand tomorrow's work if I have not mastered yesterday's. In my experience, it is much more of a spiralling, layering process, of things coming clearer with time and repeated exposure.

Sensible levels of expectation are closely linked to the activities of specializing and generalizing. On first exposure to a mathematical topic, my task is to let go of details and get an overall impression of what is going on. If I am being offered specific examples (specializations), then my task is to generalize, to work on the examples by trying to see what is common, and thus to see what is going on. If I am being offered generalizations or theory at first, then my task is to seek specializations for myself, and this requires specific examples which give me confidence. It is often very tempting to try to work through the theory line by line, but unless this is done with examples which *I* am confident in manipulating, it gives no overall sense of what is going on, and is rarely much use. Thus *I* must take responsibility for constructing examples suitable for me. As I work more closely with the examples, my aim is to try to see for myself the general principle behind them. Each attempt to articulate that generality for myself is a conjecture which is (I hope) leading me towards deeper understanding. When I feel the need to practise the technique on examples, I am laying down a firm bed of practical experience which will come back to me when called upon later.

When written out like this, it seems like a lot to have to do as well as study, but the whole point is that this *is* what studying actually means. It *is* what you are doing already. By being aware of the processes, however, you will almost certainly find that you can become more efficient, and less prey to worries about not mastering things on first exposure.

Advice

Try to establish reasonable expectations of mastery and understanding. Mastery comes with practice, understanding with time.

Do not rush too quickly into trying to master a technique until you have worked on examples and tried to write down for yourself what principles or ideas are being illustrated.

Section 4 CONVINCING YOURSELF AND OTHERS

When you think you have understood something, or found an answer to a question, it is easy to be convinced that you are right, simply because it is *your* idea. It can come as quite a shock, several weeks later, to discover that what was once so clear is now obscure or even incorrect. Your idea or solution may be right, but it may be only partly right or even completely wrong. If it is going to serve as a stepping stone in the growth of understanding, then it should be doubted and questioned.

This section is concerned with the transition from convincing yourself to convincing someone else. It is easy to get the impression from mathematics texts that arguments are supposed to spring full blown from someone's mind, when in fact quite the reverse is the case. Any argument of any significance has been through many stages of change and refinement.

The principal device being offered is to pay attention at all times to what you KNOW and what you WANT, both of which are clarified by specializing and generalizing. Every argument (mathematical not social!) consists of a bridge built between what is KNOWN at the beginning, and what is WANTED at the end. However brief or skeletal, the intervening steps are intended to indicate how to pass from the known to the unknown. Despite its simplicity, this observation can be very helpful indeed in trying to resolve mathematical questions.

Consider the following statement, which began the first section, on specializing.

> The sum of the cubes of the first N positive integers is the square of the sum of those integers.

We specialized a few cases to get a sense of what the statement was saying, but did not address the question of whether the statement is actually always correct. There are many different approaches using technical devices such as induction or binomial coefficients. The approach taken here shows a geometrical way of looking at it. It also illustrates how you might go about finding your own argument.

The first thing to decide is whether you are convinced that the statement is reasonable. This means trying some simple examples to see if they are right; but it means more. It means trying to get some sense of *why* it seems right. Just because it checks in three of four cases, we have no reason to believe that it will always work. For example, is it true for the first 37, or the first 137 positive integers? Try some more cases, and try to get a feel for what is happening.

■■DO SO NOW■■■■■■■■■■■■■■■■■■■■■■■

After three or four more cases, it becomes hard to believe that it could ever fail! This is the moment to be on guard, because it is not enough to be sure in yourself—as many mathematicians have found to their chagrin. What is needed is a convincing argument, and to find it, you must continue specializing until you have found some reason *why* it works.

In this case, I checked it for $N = 5$:

$$1^3 + 2^3 + 3^3 + 4^3 + 5^3 = 225$$

and

$$(1 + 2 + 3 + 4 + 5)^2 = 225$$

To check it for $N = 6$, I noticed that it is not necessary to repeat all the work:

$$1^3 + 2^3 + 3^3 + 4^3 + 5^3 + 6^3 = 225 + 6^3$$
$$= 441$$

and

$$(1 + 2 + 3 + 4 + 5 + 6)^2 = (15 + 6)^2$$
$$= 21^2$$
$$= 441$$

The current round of specializing is to try to see what is actually involved, and by following the *shape* of the calculations rather than just doing the arithmetic, an idea begins to emerge. Each instance could be based on the preceding case.

You might be able to see how to exploit the shape of the calculations and write down a general description of how the Nth case follows from the previous case, but my wish at this point was to be able to see directly what was happening, so I looked for a geometric picture.

I WANT $1^3 + 2^3 + 3^3 + \ldots + N^3 = (1 + 2 + 3 + \ldots + N)^2$

I KNOW very little. I have no feeling for what sums of cubes are like, but I do know that squares can be pictured as squares on paper, so I am going to go for a diagram. I want to picture N^3. A cube seems an obvious idea, but does not appeal to me because I cannot see how to pile them all up. All I can think of is N copies of an N by N square, which at least I can draw. My overall idea is to draw a diagram which shows the growth of the sum of the cubes and the square of the sum, as N increases.

Specializing

I begin by drawing a series of squares of size 1 by 1, 2 by 2, 3 by 3,.... I want them to pack into a square corresponding to the square of the sum, of size $1 + 2 + \ldots$. Try some cases:

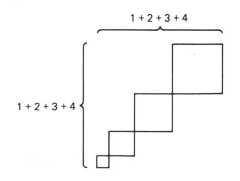

1 + 2 + 3 + 4

1 + 2 + 3 + 4

Try $N = 2$. I want one 1×1 square and two 2×2 squares.

Want   Try

AHA! The remaining rectangles are each half a 2 by 2 square.

This is a diagrammatic version of $1^3 + 2^3 = (1 + 2)^2$.

Try $N = 3$. I want one 1×1 square, two 2×2's and three 3×3's.

AHA! My three 3 by 3 squares appear.

I have $1 \times 1^2 + 2 \times 2^2 + 3 \times 3^2 = (1 + 2 + 3)^2$.

Try $N = 4$. I want to add on four 4×4 squares.

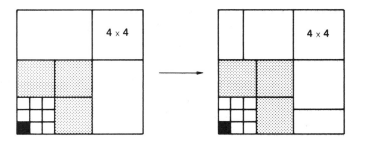

I need look only at the L-shape added on to the previous case. The two rectangles make the fourth square.

I feel ready to try a more general case now.

I KNOW that I have to fill an L-shape which is N steps wide and which has arms of length $1 + 2 + 3 + \ldots + (N - 1)$.

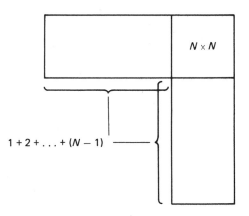

An N by N square fits in the corner. How many N by N squares fit in each arm? I expect different answers for N even and N odd.

To know the lengths of the arms, I need to know $1 + 2 + 3 + \ldots + (N - 1)$. From Exercise 3.3, it is $N(N - 1)/2$. So the two arms together constitute a strip which is $N(N - 1)$ steps long and N steps wide. Such a strip will be filled by $N - 1$ squares which are N by N, so including the one at the corner, there are N squares of size N by N to fill in the L-shape, exactly as requested. When N is even, one N by N square will appear as two rectangles.

The important feature of this example is *not* in the details of the argument, but in the helpfulness of referring frequently to what is KNOWN and what is WANTED. Initially, these are provided in the question. As work begins, more specific and detailed WANTS arise, and at the same time new relevant KNOWNS are generated.

Whenever I get stuck trying to produce a convincing argument, I find it helpful to write down everything I know that seems relevant—it rarely takes me long, and it helps to get it on paper in front of me, because otherwise my head gets too full of half-formed ideas to be able to concentrate properly.

The argument as written down is just as it occurred to me. I find it convincing because it shows how the sum of cubes and the square of the sum grow with increasing N. I no longer depend on a few examples, nor on a vague feeling that it seems to be right. I can actually see the structure in the diagrams. It is almost as if the diagrams speak—but it takes a lot of words to write out what they say. To make it convincing to others, I shall have to add more words, structured by KNOW and WANT. Why not try right now to write down in your own words and symbols what the diagrams are actually saying to you?

■■■ WRITE IT DOWN NOW ■■■■■■■■■■■■■■■■■■■

Here is my version.

Assuming I KNOW $1^3 + 2^3 + \ldots + (N - 1)^3 = (1 + 2 \ldots + (N - 1))^2$ for a particular value of N, I WANT to add N^3 to both sides of the equality and see if the right-hand side is the square of the sum of the integers from 1 to N. The diagram shows me that N^3 is the area adjoined, which is 2 copies of $N(N - 1)/2 \times N$ plus N^2 in the corner.

The right-hand side is now

$$(1 + 2 + \ldots + (N - 1))^2 + 2N^2(N - 1)/2 + N^2$$
$$= (1 + 2 + \ldots + (N - 1))^2 + 2(1 + 2 + \ldots + (N - 1))N + N^2$$
$$= (1 + 2 + \ldots + N)^2$$

This is what I WANTED.

It is not always easy to write down what you see in a diagram, which is why diagrams are so useful. The eye can take in several features simultaneously, whereas words are often limiting by virtue of their sequential nature. At the same time, diagrams have a weakness. What we think a diagram expresses may be sufficiently complicated that we are misled, and get it wrong. Mathematicians, conscious of having been misled by diagrams in the past, like to have everything written out in symbols. Nevertheless, mathematicians often depend on diagrams and images to know what to write down!

Earlier, I wrote down the arithmetical calculations for $N = 5$ and $N = 6$, and suggested that these might also speak to you in much the same way as a diagram does. In this case, an argument can be revealed by systematically refusing to do arithmetic calculations.

Specializing

Let us repeat the $N = 5$ and $N = 6$ cases. I assume that I have already shown that

$$1^3 + 2^3 + 3^3 + 4^3 + 5^3 = (1 + 2 + 3 + 4 + 5)^2$$

I WANT

$$1^3 + 2^3 + 3^3 + 4^3 + 5^3 + 6^3 = (1 + 2 + 3 + 4 + 5 + 6)^2$$

Using what I KNOW in the left-hand side,

$$1^3 + 2^3 + 3^3 + 4^3 + 5^3 + 6^3 = (1 + 2 + 3 + 4 + 5)^2 + 6^3$$

I WANT this to be the same as

$$(1 + 2 + 3 + 4 + 5 + 6)^2$$

which I KNOW to be the same as

$$(1 + 2 + 3 + 4 + 5)^2 + 2(1 + 2 + 3 + 4 + 5)6 + 6^2$$

Comparing the two expressions, I find that $(1 + 2 + 3 + 4 + 5)^2$ is common to both sides, so I WANT to work on the remainder.

The awkward term is $(1 + 2 + 3 + 4 + 5)$, which I KNOW to be $5 \times 6/2$, using the formula for summing consecutive integers.

I WANT the value of $2(1 + 2 + 3 + 4 + 5)6$ which is

$$2 \times (5 \times 6/2) \times 6 = 5 \times 6 \times 6$$

To this I must add 6^2, giving

$$5 \times 6^2 + 6^2 = 6^2(5 + 1)$$

$$= 6^3, \text{ as requested}$$

Generalizing

Now I can go through and replace 5 by $N - 1$ and 6 by N. The result of this act of generalizing will be the same as my description of the diagram.

■■■TRY IT NOW■■■

One further point which emerges from the sum of cubes is the difference between convincing yourself and convincing someone else. It is so easy to be beguiled by a conjecture, especially if it is your own. It is essential to get it written down, and then to disbelieve it. This means more specializing, in a more detached way, trying to see *why* it might be right, or how some example might show it to be false. When you have convinced yourself, then it is time to convince someone else, either real or imagined. On this next run through, you must be extremely sceptical, just as a colleague or tutor might be. You look for places where there is still a gap between a detailed WANT and a detailed KNOWN, and see if it needs bridging with more detail. The best way to learn to be sceptical is to practise on other people's arguments.

To tackle someone else's argument, you first need to get an overview of what the argument purports to demonstrate. Then, line by line, you ask yourself

What do I KNOW?

and

> What do I WANT?

in that line. Can a bridge be built, or is there a difficulty? When you get to the end, it is worth reviewing the argument to pick out the main features, the turning points, for the rest of it can be reconstructed from them.

Here is a short example from a textbook.

> The technique...can be adapted to solve any equation of the form
> $$a^x = k,$$
>
> for x, where a and k are positive real numbers and $a \neq 1$.
> Specifically, the technique is as follows.
>
> $$a^x = k.$$
>
> Apply $\log_{10}$:
>
> $$\log_{10}(a^x) = \log_{10}(k),$$
>
> then
>
> $x \log_{10}(a) = \log_{10}(k)$ ('log of a power = multiple of the log'),
>
> then
>
> $x = \log_{10}(k)/\log_{10}(a)$ (because $a \neq 1$, we know that $\log_{10}(a) \neq 0$).

When KNOW and WANT are inserted, it comes out as follows.

I WANT to find x explicitly, given $a^x = k$.

I KNOW $a^x = k$, $a \neq 1$.	I WANT x; I WANT to get x out of the index position.
I KNOW logs strip down exponents.	I WANT to apply logs to both sides.
I KNOW $a^x = k$ $\quad \log_{10}(a^x) = \log_{10}(k)$.	I WANT x.
I KNOW $\log_{10}(a^x) = x \log_{10}(a)$ (that's why I used logs!).	I WANT x.
I KNOW $x \log_{10}(a) = \log_{10}(a^x)$ $\qquad\qquad\quad = \log_{10}k$.	I WANT x.
I KNOW $x = \log_{10}(k)/\log_{10}(a)$.	Which is what I WANTED!

Having inserted all the KNOWs and WANTs, I found no lines which needed augmenting. The advantage of putting them all in is that at each stage I am reminded of where I am going—one of the hardest parts of reading mathematics is retaining a sense of where you are going, in the midst of algebraic detail.

At the end of this text extract, it says that the technique is straightforward, and it is probably easier to apply the technique from first principles each time than to try to remember the formula. What is involved in remembering the technique? Looking back over the argument, the key features are the use of logarithms to strip off the exponential, and keeping track of what you want, namely to solve for x.

The argument can now be collapsed to

solve $a^x = k$ for x,

take logs of both sides, and

$x = \log_{10}(k)/\log_{10}(a)$.

This very truncated version can be expanded using KNOW and WANT, whenever this is needed.

Here now is an example for you to try, with my resolution, which you should look at only after you have made an attempt. It is too easy to read my version and then say afterwards, 'oh yes, I could have done that', or even 'I could never do that!'. The only way to strengthen your mathematical muscles is to exercise them, on problems as well as texts.

Divisors

Is there a number with exactly fifteen divisors including 1 and itself?

■■■TRY IT NOW■■■■■■■■■■■■■■■■■■■■■■■■■■■■■■■■■■■■■

Resolution

Fifteen is far too large. What shall I do? I could flail around looking for a number that works; I could simplify 15 and look for numbers with just 1 divisor, then 2, then…; I could look at the number of divisors of 1, 2, 3, 4,…in turn. I shall specialize down to 1, 2…divisors. I must be clear about what divisors are, so try some examples,

| Specializing |

1 has 1 divisor, namely 1.

2 has 2 divisors, namely 1 and 2.

3 has 2 divisors, namely 1 and 3.

4 has 3 divisors, namely 1, 2 and 4.

I wrote down the number of divisors of 5, 6, 7, 8, 9, 10, 11 and 12, and I recommend you do too. I went as far as 12, because I know 12 has more divisors than the earlier numbers.

■■■DO SO NOW■■■■■■■■■■■■■■■■■■■■■■■■■■■■■■■■■■■■■

Now, what do I WANT?

I want a number with 15 divisors. It is going to take a long time to build up to that, so how can I see what is really going on? What do I KNOW about divisors?

| Generalizing |

If I am looking at divisors of N, then they come in pairs, because if a divides N, so does N/a. OOPS! If N is a perfect square, then its square root has no mate. But that is the only way that a divisor would not be paired up. I have reached a pretty strong conjecture.

CONJECTURE: only perfect squares have an odd number of divisors.

I checked it on the examples I had already done. It seems right in those particular cases, and my argument seems OK. I am pretty satisfied with the conjecture, so I shall carry on. I WANT a number with 15 divisors, so it must be a perfect square. Because it is a perfect square, one of the divisors is accounted for, but where will the other 14 come from? More systematic specializing is needed with perfect squares!

1 has 1 divisor.
4 has 3 divisors.
9 has 3 divisors.
16 has 5 divisors.
25 has 3 divisors.
36 has 9 divisors.

Notice that the specializing is slightly different now, as a result of the conjecture. What about 7 divisors? Can that ever happen? Try some more:

49 has 3 divisors.
64 has 7 divisors.

Now I begin to see a pattern. Look at the divisors of 64 clustered either side of 8:

1   64
2   32
4   16
    8

I conjecture that to get 15 divisors, I should take 2 raised to the $(15 - 1)$, since one divisor comes from the square root, and the other divisors are powers of 2, coming in pairs.
    CHECKING: The divisors of 2^{14} are

$$1, 2, 2^2, \ldots, 2^{11}, 2^{12}, 2^{13}, 2^{14}$$

It is important to reflect on what has happened, so that I can learn from the experience. My immediate feeling on writing down the powers of 2 was that somehow I 'ought' to have seen that right away. What seems obvious now was not obvious when I started! I have accomplished a good deal more than was asked, however, so it is useful to be clear on just what I have found out! At the same time, I want to write down a convincing argument. All that I have written so far is a record of my thoughts. Now I massage it into presentable mathematics!

Conjecture 1: A number N has an odd number of divisors if, and only if, it is a perfect square.

Argument: If I KNOW that N is a perfect square, then all of its divisors come in pairs except for the square root of N, so N has an odd number of divisors. If I KNOW that N has an odd number of divisors, then when the divisors are paired up, there must be one divisor left over which pairs with itself. It must be the square root of N, so N is indeed a perfect square.

Conjecture 2: For any odd number N, there is a number with exactly N divisors.

Argument: $2^{(N-1)}$ has exactly N divisors.

I have not used the fact that I KNOW N is odd, anywhere in the argument. This suggests the following.

Conjecture 3: For any number N, there is a number with exactly N divisors.

Argument: $2^{(N-1)}$ has exactly N divisors.

What is the special role of 2 here?

| Generalizing |

> Conjecture 4: There are infinitely many numbers with exactly N divisors.
>
> Argument: For any prime number p, $p^{(N-1)}$ has N divisors.

Now I want to look at a general number and see how many divisors it has. For example what happens if I take two primes, p and q, and form

$p^{(S-1)}q^{(T-1)}$? How many divisors does it have?

TRY IT NOW

There are S choices for powers of p, and T choices for powers of q, so there are ST choices all together. So I could get 15 divisors from any number of the form

$$p^{3-1} \times q^{5-1}$$

| Specializing |

Taking $p = 3$, $q = 2$ gives me

$$3^2 \times 2^4 = 9 \times 16$$

$$= 144$$

which is considerably smaller than $2^{14} = 16\,384$. I am struck by how much more helpful the notation of $3^2 \times 2^4$ is than the decimal notation 144.

| Generalizing |

Now I find myself posing the new question: what is the smallest number with exactly N divisors? Which is likely to be smaller, $3^2 \times 2^4$ or $2^2 \times 3^4$? What does this suggest in general?

TRY IT NOW

I have an outline idea—factor N and use a generalization of the $p^{S-1} \times q^{T-1}$ idea.

What is the value in tackling a question like **Divisors?** I suggest that unless you pause and reflect on what actually happened, there is very little value at all! Now that you are familiar with the mathematical content of the question, you can look back over what happened and learn from the experience. I suggest that, in particular, you review the resolution above, paying particular attention to the way in which frequent clarification of KNOW and WANT enabled a bridge to be seen between them.

REVIEW IT NOW

It is often the case that as you specialize, what you KNOW actually changes, and bearing constantly in mind what you WANT brings about subsidiary aims or new WANTs. Frequently, students and researchers become so caught up in their specializing and in what they KNOW, that they lose sight of what it is they really WANT. That is why I have been stressing both KNOW and WANT at the same time. By keeping a balance between them, devoting some attention to each, a sense of direction emerges. Furthermore, any resolution will consist of a bridge between KNOW and WANT. The act of setting down a convincing argument involves retracing your steps over all the intervening KNOWs and WANTs, constructing a sequence of little bridges, rather like a medieval beam bridge.

Summary

A convincing argument consists of a bridge built between what is KNOWN at the beginning, and what is WANTED. An incomplete or confusing passage in a text can be clarified by setting out what is known and wanted at each stage. An investigation can be directed by paying attention to what is known and wanted, and by constantly modifying or rephrasing them, looking for a bridge. Finally, when you think you have a convincing argument, it can be structured in terms of KNOW and WANT to help your reader see all the steps. The only way to get KNOW and WANT working efficiently for you, is practice—the next section provides plenty!

Exercises

4.1   What is the smallest number with exactly 12 divisors?

4.2   In Exercise 3.1, there was a conjecture that any acute-angled triangle can form the four congruent faces of a tetrahedron, but that it does not work for obtuse or right-angled triangles. Scissors and paper provided evidence, but a convincing argument is needed. Construct one!

STUCK? Somehow, you have to get a grip on a general triangle, and literally construct a tetrahedron, imposing the condition that all the faces need to be congruent. A three-dimensional object is hard to work with, so coordinates might suggest themselves—but they must be carefully chosen. Before resorting to coordinates, note that you WANT something about obtuse and acute angles, so focus on angle sizes. You KNOW that the tetrahedron can be folded together from four triangles. Specialize to the case when two of the flaps just meet when folded over flat. What do you KNOW about flaps that just meet, or fail to meet?

If you do resort to coordinates, try putting a general triangle in the plane with the third coordinate zero, but with as many other cooordinates zero as possible. It might help to begin with a specific numerically-coordinated triangle, and try to construct the tetrahedron. This will indicate what calculations to do with symbols. Don't forget that you WANT to end up with a condition on the original triangle which discriminates between obtuse- and acute-angled triangles.

4.3   By investigating the sequence of numbers

$$\sqrt{2}, \sqrt{2^{\sqrt{2}}}, (\sqrt{2^{\sqrt{2}}})^{\sqrt{2}}, \ldots$$

show that it is possible to find an irrational number which can be raised to an irrational power, and yet yield a rational number as the result. Generalize.

4.4   Given any two rational numbers, can you find a rational number in between them which has a power of ten as its denominator?

Given any two irrational numbers can you find a rational number in between?

Given any two rational numbers, can you find a irrational number in between?

STUCK? What is a rational number? What is an irrational number? Try some examples. How can rationals best be represented—as fractions, decimals, ...?

4.5 If $A/B < C/D$, what can be said about the sizes of $(A + C)/(B + D)$ and $(2A + 3C)/(2B + 3D)$ compared to A/B and C/D?

Place this question in a more general context by constructing a story about a journey, with different average speeds during different phases. Place it in a more general context by drawing a parallelogram with vertices at the origin, the points (B, A) and (D, C), and the point $(B + D, A + C)$.

STUCK? Specialize, then generalize.

Interlude D # ON BUILDING CONFIDENCE

Lack of confidence is one of the main features of being a student. For me, very often it turned out that I actually knew a great deal more than I thought, but I lacked confidence in myself, and in my ability to get started on a question. What is the basis of mathematical confidence?

Here is a list of increasingly abstract mathematical ideas, at least some of which you have met before:

> the integer 37
> the rational number $\frac{4}{5}$
> the real number $\sqrt{2}$
> the real number π
> the real number e
> the real number x
> the point (x, y) in the plane
> $\{(x, y): y = 3x + 2\}$
> $\{(x, y): y = mx + c\}$
> the function $x \mapsto x^2$
> the function $\sin: \mathbb{R} \mapsto \mathbb{R}$
> the function $f: \mathbb{R} \mapsto \mathbb{R}$
> $\{(x, y): y = f(x)\}$
> $\{F: F \text{ a function } \mathbb{R} \mapsto \mathbb{R} \text{ and } F(0) = 0\}$.

I expect that you feel confident that you can manipulate some of these, but not others. For some, you will have a sense of what they are about, while others may be more or less mysterious. But how many of them would have inspired your confidence before you began studying? I suspect that many more would have belonged to the nonsense category than do now. You *have* made progress!

Starting with manipulating objects, which may be physical objects, or more abstract mathematical gadgets like the ones listed, I gradually develop a sense of a more complex notion, and then try to capture it in some sort of picture, words and notation. With experience and practice, the symbols become more succinct, and at the same time more confidence-inspiring. With mastery, they turn into confidence-inspiring objects which I can use in further specializations to build yet more complex notions, such as those near the bottom of the list.

One of the characteristics of mathematical thinking is the number of times I traverse a spiral from

> confidently manipulable objects

through

a sense of a general notion common to the specializations

through

an increasingly succinct notation for the generality, in pictures, words and symbols

to

new confidently manipulable objects.

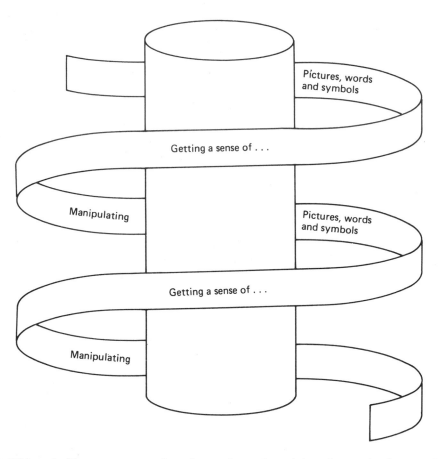

This spiralling goes on and on in mathematics, giving rise to the frequently misunderstood idea that mathematics is a subject in which you cannot understand a new idea unless you have understood everything that has gone before. Instead of an all-or-nothing 'have you got it or haven't you', my experience is much more like layers of earth being laid one on top of another, ultimately providing soil for germinating new seeds. What at first seems highly abstract will, with time and experience, become familiar and confidence-inspiring. The particular collection of objects which I can confidently manipulate is my base. Yours is probably different. We can both tackle a new idea, but we shall use different examples on which to specialize. Consequently, the way we speak about what is going on is likely to be different, at least at first.

Transition from being able to manipulate examples to having a sense of what is actually going on involves specializing and generalizing, which the main sections of this unit have described in detail. Moving from a vague sense of something which is pre-verbal, to being able to capture that 'sense of', involves taking time over trying to say what it is that I see, and then recording it in words, in mixtures of words and symbols, and finally in succinct symbols. Even that is not the end of the story, however, because the symbolic version is of no particular value unless it speaks to me directly, and unless I can confidently manipulate those symbols. To reach this state takes time and practice. Instead of rushing into symbols, it is very sensible to stick to versions that I do feel happy with. This often means translating other people's symbols into my own words and pictures, and that is precisely how any mathematician deals with new material. Put another way, it is virtually impossible to *read* mathematics—it has to be *done*, to be worked through with pencil and paper.

Being aware of the spiral of

> manipulating,
> getting a sense of,
> capturing in symbols,
> fodder for further manipulation,
> ...,

helps me to keep my expectations at a reasonable level, and to specify clearly what mathematical activities are required by the immediate study task. Students who expect to jump from exposure to mastery are liable to miss out the essential intermediate state of transforming experience with special cases into their own generalizations. The whole purpose for recommending that you spend time

- specializing —manipulating confidence inspiring objects
  —trying to get a sense of a pattern or idea

- generalizing—trying to talk about that 'sense of'
  —trying to capture the pattern or idea succinctly

- exercising —practising examples and techniques to reach mastery so that the succinct articulation becomes in its own time a confidently-manipulable object

is so that when you encounter a difficulty, you can track back down the spiral. You will then have a variety of written, pictorial and verbal descriptions which contribute to your growing sense of the topic at hand. Because *you* have articulated them, they are yours, and they can be accessed. If they simply come from a book, then they are somebody else's!

Advice

Abstract symbols will become concrete and confidence-inspiring if they are succinct summaries of ideas that you are confident with. Don't rush into using symbols, but don't be afraid of them.

When a text or question seems overwhelming in its use of symbols or complex terms, don't panic! Backtrack, rewriting them in a form that gives you confidence—specializing—and then work your way back to the substance of the passage—generalizing.

At the end of each study session, and each text section, write down your own summary of what it was about—the main ideas, the techniques, and the sticky places. Try rehearsing them to yourself while waiting for a bus or at other odd moments during the day. If possible, compare notes with other students, or with your tutor.

Section 5 WHEN IS AN ARGUMENT VALID?

Since doing mathematics is concerned with investigating generalizations and convincing both yourself and others that the generalizations are valid, it would be nice to be able to say definitely what constitutes a valid and totally convincing argument. Unfortunately, this is rather difficult. In order to be mathematically precise, it would be necessary to provide minute details justifying every little step. Even then there might be some assumption hidden in the implicit meanings of the words. To be absolutely certain, the words have to be converted into symbols which are manipulated mechanically according to formal rules. All sense of meaning is abandoned. Such a proof of anything interesting would be so cumbersome as to be uncheckable, and certainly it would be uninformative as to *why* the proof was valid and the statement true.

Mathematicians (apart from those interested in the theory of proofs) are much too interested in the investigation itself to be bothered with all the fine details which only get in the way and obscure the overall picture, so the detailed steps are omitted. Usually this is perfectly safe—but occasionally the steps become leaps over unnoticed chasms. Indeed, it has been suggested that despite rigorous refereeing of mathematical journals, most research articles contain errors, and a high proportion contain serious errors or omissions. These have a habit of turning up later, sometimes much later. Many of these errors can be attributed to the high costs of paper, and hence the demand for succinct outlines of proofs, but it also happens that subtleties are overlooked. If mathematicians have trouble with their arguments, it is not surprising that students do too.

This section outlines a few examples of oversights which have been made in the past, but which have later led to fruitful mathematical investigations. The bulk of the section is devoted to examples of partial and erroneous arguments for you to examine critically.

Famous Oversights

The *Four-colour Theorem* says that for any map drawn on a flat piece of paper, the countries can be coloured using only four colours, in such a way that no two countries with common boundaries have the same colour. The first 'proof' appeared in 1879, and was accepted for some 11 years as valid, before being shown to have a missing step that no one could bridge. The missing step turned up when someone tried to generalize the argument to maps on different surfaces, like an inner-tube, and found that the method could be adapted to work in all cases *except* the plane! It was only in 1976 that a convincing argument was found for the planar case, but it involved many hours of computer calculations.

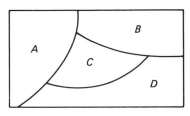

One point that arises from this example is that a general statement can be shown to be inadequate by exhibiting just *one* case, one specialization which is false—often referred to as a counter-example. Thus the statement

all primes are odd,

is shown to be false by observing that 2 is not an odd prime. In this case, it is the *only* counter-example! To show that an argument is *not* valid can be a lot trickier. If the statement it purports to justify is itself false, then of course the argument must be false. In the case of a result that turns out to be true, such as the four-colour theorem, errors in arguments cannot be found by producing counter-examples. The place where an unjustified leap occurs must be found. Mathematics departments in universities frequently receive documents purporting to prove famous outstanding conjectures, and each must be seriously read, because they might conceivably be correct. By way of contrast, arguments that purport to show how to trisect an angle using ruler and compasses only, are known to be in error because what they try to do has already been shown to be false.

Another interesting historical example is given by *Steiner's* argument to justify the statement

the largest region which can be surrounded by a stick, with a rope whose ends are tied to the ends of the stick, is a segment of a circle.

The thrust of the argument is that if the stick and the rope do not form a circular segment, then a construction can be given for encompassing a slightly larger region. The only configuration which can *not* be made larger by this construction is a circular segment, therefore a segment of a circle gives the largest area.

The details of the construction are not important, for the subtlety lies in the assumption that the process of making the area larger and larger must converge to some region which cannot be made larger by this construction. Intuitively the argument seems reasonable enough, but there is a similar argument which indicates what might go wrong.

CONJECTURE: One is the largest whole number.

Argument: Any whole number can be made larger by squaring it, except for 1 which stays the same. By the same argument as in *Steiner's* circular segments, keep on making the number bigger by squaring it, and since 1 is the only number that stays fixed by this construction, you must eventually get close to 1. Thus 1 must be the largest.

Steiner's argument *is* valid, whereas this is not, but the same strategy is used. Where does the analogy break down?

▉ TRY IT NOW ▉▉▉▉▉▉▉▉▉▉▉▉▉▉▉▉▉▉

(My resolution is on page 78.)

Intuition about when things converge needs to be developed and tested, because our naive intuition is not always trustworthy.

An infinite sum such as

$$1 - \tfrac{1}{2} + \tfrac{1}{3} - \tfrac{1}{4} + \tfrac{1}{5} - \tfrac{1}{6} + \cdots$$

certainly looks as though it will converge, because as you start to calculate from the front, it begins at 1, then goes down to 1/2, then up by only 1/3, then down by only 1/4,.... It appears to be converging to some number between 1/2 and 1 (in fact it converges to 1). But if you rearrange it slightly as

$$(1 + \tfrac{1}{3} + \tfrac{1}{5} - \tfrac{1}{2}) + (\tfrac{1}{7} + \tfrac{1}{9} - \tfrac{1}{4}) + (\tfrac{1}{11} + \tfrac{1}{13} - \tfrac{1}{6}) + \cdots$$

in which each bracket is formed by taking enough positive terms to exceed the next unused negative term, then each bracket is positive, the first is bigger than 1, and every term of the sum is used up eventually. This rearrangement appears to converge to something bigger than one. Rather than throw up our hands and rule out infinite sums altogether (which would make us lose decimal fractions like 0.3, $\sqrt{2}$ and π, and so render decimal numbers pretty useless), it is sensible to examine in more detail when rearrangements make a difference and when they do not, and then to use that theory to educate our intuition. Surprises and unexpected difficulties should all be treated as opportunities to extend and correct your sense of how things fit together. The definitions and theorems of mathematics are aids to refine and develop intuition, not things to be memorized.

Very often, the existence of an undetected subtlety emerges when someone appears to be excessively pedantic in their question of hidden assumptions, and as a result, new areas are opened for mathematical investigation. Here are two examples.

It seems quite clear that if you pick up a pencil and draw a continuous curve which never intersects itself, finally ending up by joining on to where you started, then what you have done is to divide the plane piece of paper into two regions—an inside and an outside.

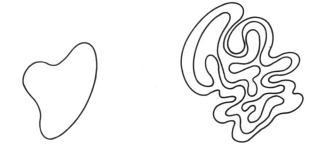

Unfortunately, it is not always clear in a complicated drawing just which is the inside and which the outside. When you try to write down an argument to justify this seemingly innocent and 'obvious' assumption, it turns out to be harder than it looks. Indeed, it requires a great deal of mathematical machinery in order to give a convincing proof, because lying behind the difficulties are several unstated assumptions about what constitutes a plane, a continuous curve, and a region. Investigation of this question, once it was exposed as being not completely obvious, opened the way to the whole subject of topology, which lies at the heart of modern mathematics.

To count a finite number of objects is to utter a 'counting poem' while picking out each object in turn. By analogy, to count an infinite set of 'objects' is to assign to each object, one of the members of a counting set ('counting poem') such as 1, 2, 3, 4.... But this method of counting leads to some surprises. The even numbers can be counted by the set 1, 2, 3,... of which they form only a part—as each even number comes by, announce whether it is the first, second, etc.! Indeed the same thing can be done with the square numbers, the cubes, the numbers which leave a remainder of 3 on dividing by 17, and so on. Our intuition of finite numbers tells us that 'bigger' means 'larger count' as well as 'cannot be a subset of'. For infinite sets, the subset sense of 'bigger' is no longer appropriate because it leads nowhere. The 'larger count' approach proves to be very fruitful. The whole question of what constitutes an infinite set took mathematicians a long time to deal with, and is not a simple matter to be skipped over lightly.

While trying to clarify what sets and subsets one could talk about sensibly, Bertrand Russell discovered that the problem is tied up with self reference, which with the advent of powerful computers has begun to be tamed under the name 'recursion'. Russell's discovery is best illustrated by his adjective paradox. Since some words, like 'short', actually describe themselves, while others, like 'monosyllabic', do not, he proposed calling words that fail to describe themselves 'heterological'. The question then arises as to whether 'heterological' is itself heterological—that is, whether it describes itself or not.

Which ever way you decide, you reach a contradiction. Thus if 'heterological' is itself heterological, then it fails to describe itself. But 'failing to describe itself' is exactly what heterological means, so it does describe itself—a contradiction! Investigate the result of taking 'heterological' as not being heterological.

■■■■TRY IT NOW■■■■

Our intuitive ideas about forming sets and making choices break down in language and in mathematics, until they are refined, and made more mathematically precise. See the last page of this book for more examples.

Looking for Omissions

If great mathematicians overlook steps in their arguments, what hope is there for lesser mortals? The short answer is that there is no foolproof antidote, but there are steps to be taken to exercise care, and to improve your mathematical thinking. They are the following.

1. Always be clear about the status of a statement—conjectured or proved. Remember that specialization is suggestive, and can *indicate* why

something might be true, but does not constitute a proof unless all possible specializations have been tested.

2. Always state clearly what you are aware of assuming. One good way to do this is to state clearly what you KNOW at each stage.

3. Keep clear what you KNOW and what you WANT. If you modify what you WANT, remember to check later whether you have really answered the original question.

The remainder of this section consists of examples of valid and invalid arguments for you to examine critically.

Examples

(My comments begin on page 78.)

5.1   The function $x \mapsto (\sqrt{1 - x^2})(\sqrt{x^2 - 4})$ never exceeds 2.

Argument: the first square-root term is never bigger than 1, and the second never bigger than 2.

5.2   The function $x \mapsto \log(\log(\sin(x)))$ has no maximum value.

Argument: log is an increasing function.

5.3   If $(x - 2)^2 + (y + 3)^2 = 0$, then $x = 2$ and $y = -3$.

Argument: The square of a real number is never negative.

5.4   The function $F: x \mapsto x^3 + 2x + 1$ has no zeros when x is an integer and all calculations are done in arithmetic modulo 3.

Argument: $F(0) = 1$, $F(1) = 1$, $F(2) = 1$.

5.5   In a certain university, there are 14 times as many students as lecturers. The equation which represents this is $14S = L$.

Argument: Translate the phase '14 times as many students as lecturers', word by word.

5.6   The number $2 \times \pi^{\sqrt{163}}$ is an integer.

Argument: My calculator gives it as $4\,448\,511$.

5.7   The inner polygon in the figure at the top of page 55 is a regular octagon.

Argument: The figure is symmetrical about the line AE, so AB is equal to AH.
The figure is symmetrical about the line DH, so AH is equal to HG. The figure remains the same if rotated through 90° about its centre, so $HG = GF$. Thus all edges of the inner polygon are equal in length, so the octagon is regular.

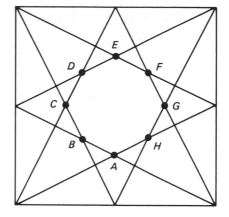

5.8 If A, B, X and Y are positive real numbers, then

$$\sqrt{A+B} \times \sqrt{X+Y} \geqslant \sqrt{AX} + \sqrt{BY}.$$

Argument: Square both sides to get

$$(A+B)(X+Y) \qquad \geqslant AX + BY + 2\sqrt{AXBY}$$

$$AX + BX + AY + BY \quad \geqslant AX + BY + 2\sqrt{AXBY}$$

$$AY + BX \qquad \geqslant 2\sqrt{AXBY}$$

$$(AY+BX)^2 \qquad \geqslant 4(AXBY)$$

$$A^2Y^2 + 2AYBX + B^2X^2 \geqslant 4AXBY$$

$$A^2Y^2 - 2AYBX + B^2X^2 \geqslant 0$$

$$(AY-BX)^2 \qquad \geqslant 0, \quad \text{which is true.}$$

5.9 If you take K consecutive numbers and add them, then exactly one of the K consecutive numbers *or* their sum is divisible by $K + 1$.

Argument: For $K = 2$ it is clear, because if the two consecutive numbers are N and $N + 1$, then their sum is $2N + 1$. Then N, $N + 1$ or $2N + 1$, is divisible by three, but in each case neither of the others can be. The argument for $K > 2$ proceeds in the same fashion.

5.10 There is no difference between tall people and short people.

Argument: A seven-foot person is certainly tall, and a four-foot person is certainly short. There is no visible difference between people who differ by 0.1 of an inch. Therefore there is no difference

between people who are 4 feet, and people who are 4 feet 0.1 inch;
between people who are 4 feet 0.1 inch and people who are 4 feet 0.2 inch;
between people who are 6 feet 11.9 inches and people who are 7 feet.

5.11 The maximum number of regions that can be formed in a circle by drawing all possible chords between P points on the circumference is 2^{P-1}.

Argument:

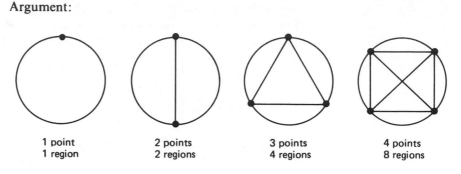

| 1 point | 2 points | 3 points | 4 points |
| 1 region | 2 regions | 4 regions | 8 regions |

Clearly, the pattern continues, doubling each time.

5.12 $\sqrt{2}$ is irrational.

Argument: Either $\sqrt{2}$ is rational or it is irrational. Suppose it is rational, so that $\sqrt{2} = p/q$, where p and q are integers; then $2q^2 = p^2$. Now the square of any number, when written as the product of primes, must have an even number of prime factors. But the left-hand side has an odd number of prime factors, but the right-hand side has an even number, which is not possible, so $\sqrt{2}$ must be irrational.

5.13 If $A + B\sqrt{2} = C + D\sqrt{2}$ where A, B, C, D are integers, then $A = C$ and $B = D$.

Argument: $A + B\sqrt{2} = C + D\sqrt{2}$ implies $\sqrt{2}$ is a rational number unless $B = D$, in which case $A = C$.

The same argument generalizes to A, B, C and D rational numbers, and indeed real numbers, and $\sqrt{2}$ can also be replaced by $\sqrt{N}$ for any integer N.

5.14 The sum of any two even numbers is even.

Argument: $2N$ represents **any** even number, so $2N + 2N$ is the sum of any two even numbers, and that is $4N$, which is certainly even.

5.15 The sum of any two numbers is even.

Argument: N represents any number, so $N + N$ is the sum of any two numbers, and that is $2N$, which is certainly even.

5.16 One solution of the question $\cos(\sin(x)) = x$, is $x = 1$.

Argument: My calculator gives $\cos(\sin(1)) = 1$.

5.17 $\pi = 2$.

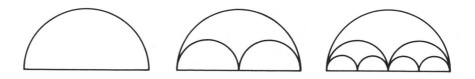

Argument: In the figure, the diameter of the semi-circle is 2, and the perimeter of the semi-circle is π. Replace the semi-circle by two semi-circles half the size of the first. Their total perimeter is still π. Replace them both by two semi-circles half their size, for which the total perimeter is again π. The arcs get closer and closer to the diameter. Eventually there will be so many semi-circles that they are virtually the same as the diameter. In the limit, the total perimeter must be the same as the diagonal, so $\pi = 2$.

5.18 Any solution in integers of $A^2 + B^2 = C^2$ must have the form of a multiple of the triple $X^2 - Y^2$, $2XY$, $X^2 + Y^2$ for some integers X and Y.

Argument: To begin, I know that

$$A^2 + B^2 = C^2$$

and I want A, B and C all to be integers. I also want to be able to produce particular instances without having to guess, so what I really need to know is what relationships must exist between A, B and C apart from the given one. Since I have only one expression to work on, I try rearranging it:

$$A^2 = C^2 - B^2$$

This immediately looks hopeful, because I am working with integers, which suggests factorizing, and I recognize that the right-hand side does factorize:

$$A^2 = (C - B)(C + B)$$

Now I want to know what I can deduce about C and B from the fact that $C - B$ and $C + B$, when multiplied together, give a perfect square. It must be that

$C + B$ is itself a perfect square times a number,

and in order for $(C + B)(C - B)$ to be a perfect square,

$C - B$ is a perfect square times the same number.

More formally,

$$C + B = X^2 \times Z$$
$$C - B = Y^2 \times Z$$

and the same Z occurs in both because it is needed to make a perfect square for A^2.

Now, what do I want? I want to find relationships between A, B and C, and I know that $C + B$ and $C - B$ must have certain shapes expressed by X, Y and Z. I can solve for C and B in terms of X, Y and Z, then solve for A.

Some calculation gives

$$C = Z(X^2 + Y^2)/2$$
$$B = Z(X^2 - Y^2)/2$$
$$A = XYZ$$

What can be made of this? I notice that Z is a common factor for all three of A, B and C, so I can omit Z, and remember that I can always multiply each of A, B and C by any factor I like.

At the moment, C and B may not be integers, because they are both divided by 2. Why not get rid of that by using the Z! Put $Z = 2$ to get:

$$C = X^2 + Y^2$$

$$B = X^2 - Y^2$$

$$A = 2XY$$

Having checked over my reasoning to look for slips, I now look back to see what I have learned...

Have I shown that *every* integer solution of

$$A^2 + B^2 = C^2$$

must have the form of a multiple of the triple

$$X^2 - Y^2, \quad 2XY, \quad X^2 + Y^2?$$

Summary

Specializing and generalizing are available whenever you feel stuck. Being clear about what you KNOW and what you WANT can clarify the way ahead in an investigation, and can explain the giant leaps in a text. The growth of mathematical understanding comes from seeing connections and modifying intuitions as a result of specializing, generalizing and convincing. It is supported by a carefully cultivated atmosphere of conjecturing, and most importantly of all, it is the source of enormous pleasure.

Interlude E   # ON EXAMPLES

Mathematics texts consist of definitions, techniques, results with proofs, and examples. What is the role of examples, and what are we really expected to do with them? Initially, this might seem a surprising question: since texts are littered with them, mathematicians must consider them important, and so the task of a student must be to learn what is in the text—or is it quite like that?

The words 'learning' and 'studying' are highly ambiguous. In one sense, 'learn' and 'study' can be interpreted to mean reading and re-reading until somehow the words and their meanings are inside me. They can also be interpreted to mean reconstructing the ideas in my own terms inside me, using the text for guidance. These interpretations can lead to very different activities.

One of the features of mathematical texts which makes them difficult to study is that examples are used for several purposes, but it is not always clear to the novice what those roles are. Hence there is a tendency either to treat them as distractions and pay them little attention, or to try to 'learn' them like learning multiplication tables, and to treat all of the text as one large body of 'knowledge' to be mastered.

Definitions, techniques and results tend to be treated identically, as 'things to be learned', whereas the authors hope that the connections *between* examples, techniques and results are what readers will concentrate on, for it is the connections which contribute to understanding.

Techniques are things you can actually do. They are procedures which are used to answer standard types of questions, such as the bisection process for solving equations. The steps can be learned by memorizing, by practice, and through understanding what they do and how they do it. Techniques are the easiest things to test, and when they have been mastered they give a sense of confidence. But it is one thing to have a tool, and quite another to know when and how to use it. Examples are essential for this.

Results (also known as theorems and propositions) and their proofs are not simply the 'facts' of mathematics. Some of them are merely technical results that are handy for use in proving other results, whereas the important ones are carefully-honed generalizations that mathematicians have worked over and over. They usually begin life as conjectures, vaguely stated, based on patterns and a sense of how things fit together. After some, and perhaps many modifications they finally emerge in the form found in textbooks, with supporting arguments. If mathematicians have struggled to express and refine their perceptions, it is unreasonable to expect to be able to appreciate the result without at least some effort. This is, of course, where examples come in again.

Definitions are the building blocks of mathematics—presumably, therefore, they ought to come first, the way they tend to in books. For example, you cannot go far in mathematics without a clear idea of what a real number is, and what a function is. However, it took decades to move from a sense of what we now call a function, to a formal definition of it, and to some idea of its ramifications. The same is true of the real numbers, and of most other ideas. In fact, definitions most commonly arise in the middle of *doing mathematics*, as an attempt to reach a succinct formulation of some conjecture or some proof.

Examples play a key role in coming to grips with techniques, results, proofs and definitions. Examples (and exercises) are used to illustrate the steps of a technique—they act as specializations of the general process. To learn the technique effectively is to reconstruct my own version, by generalizing from the examples, guided by the general exposition. To know when a technique is applicable involves recognizing the sorts of questions which it answers. Again, examples provide the specializations which I must generalize for myself. If the examples all come from a narrow context, for example in the case of bisection if all the examples are polynomial equations, then I shall associate the technique with a narrow domain of application. If I appreciate the principle behind the technique, then the technique is more likely to surface when I need it in new contexts. In the case of bisection, equations involving trigonometric functions, logarithms and exponentials (to name only a few) are all amenable to the same method. If bisection as a process is appreciated, then whenever an equation has to be solved, there is a chance that bisection will be thought of as a possibility. It is a little like recognizing faces. It is often hard to recognize people I have met only briefly, when I meet them again in a new context. So, too, it is harder to 'think' of using a technique in a new context if I have only a passing acquaintance with how it works and why.

In order for some piece of text to be an example, it must be an example *of* something. The author has in mind some general principle or theory, such as rules for manipulating inequalities based on a real feeling for numbers, their magnitudes, and a geometrical sense of the number line. Some examples are given, which for the author is an act of specializing. The first time I read them they are not yet examples *of* anything, but rather experiences of which some sense must be made, requiring acts of generalization. Of course, generalizing by its very nature is uncertain—I may stress features that the author was ignoring, and vice versa. What seem like generalizations may turn out to be variations of some but not all the examples, and generalizations may turn out to be false statements, or not the ones the author has in mind. As a reader, I am constantly looking for guidance as to which features are to be stressed, and which ignored, in order to come to what the author has in mind.

Sometimes, examples are used to give an introductory flavour of a topic, to illustrate the kinds of questions involved and to make the topic seem interesting. These sorts of examples should be carefully distinguished from other kinds, because they are not intended to be learned in the sense of memorized or mastered. They are only intended to set the scene, focus attention, and provide a flavour of what is to come. They are offered as part of an initial 'see it go by'. When the exposition begins in earnest, examples tend to be much simpler, cut to the bone so that extraneous detail does not get in the way of the reader's generalizing.

Examples play another role which is similar to, but not always the same as, the ones described so far. Whenever a mathematician is confronted by a general statement in a familiar area, there is an immediate response to specialize, to try it out on examples which have proved to be useful for this purpose. Sometimes, there is no single example, but a collection of examples. These familiar friends are the mathematician's touchstone, and have to be mastered inside and out so that they are available for confident manipulative use. When trying to recall a result, one way is to recall the examples, which then resonate with past experience and permit the result to be reconstructed.

Advice

Each time you meet an example in a text, ask yourself what it is exemplifying.

- Is it introductory?—in which case, read it for flavour.

- Is it demonstrating how a technique is carried out?—in which case, follow it through, paying attention to how it exemplifies the general technique, how it works, and what it does.

- Is it illustrating the meaning of a result?—in which case, follow it through with the general result in mind.

- Is it typical of the examples of a definition?—in which case, get to know it thoroughly.

Further Reading

1. Imre Lakatos, *Proof and Refutations*, Cambridge University Press (1976). A famous mathematical theorem is gradually clarified by means of a conversation in which arguments and objections are put forward by various characters linked to the historical development of the ideas.
2. George Polya, *Mathematical Discovery*: *On Understanding, Learning and Teaching Problem Solving*, Wiley, Combined Edition (1981).

An expensive reprint, but by far the most important and useful book on mathematical problem solving. It investigates specializing and generalizing in depth with hundreds of examples.

3. John Mason, Leone Burton and Kaye Stacey, *Thinking Mathematically*, Addison Wesley (1982).
Starting with specializing and generalizing, a framework is developed for improving mathematical thinking by learning from experience—complete with a psychological theory of how to learn from experience, and over a hundred problems to think about.
4. Philip J. Davis and Reuben Hersh, *The Mathematical Experience*, Harvester (1981).
An excellent discussion that all students of mathematics should read.
5. Susan Pirie, *Mathematics Investigations in the Classroom*, Macmillan Education, 1987.
Discusses techniques for getting pupils working investigatively.
6. Alan Graham, *Investigations in Everyday Maths*, Edward Arnold, 1987.
Intended for use with GCSE lower-grade pupils, putting mathematics in everyday context.

RESOLUTIONS

Section 1 Specializing

1.1 The squares that I have available for numbers in the 30s are 1, 4, 9, 16, 25 and possibly 36. Can I use the same square twice? It doesn't say. Try 31, 32...in sequence. I intend to subtract off a square, and see what can be done with the left-overs.

$31 = 25 + 6 = 25 + 4 + 1 + 1$, four squares if repeats are permitted.

$31 = 16 + 15 = 16 + 9 + 6$
$\qquad\qquad = 16 + 4 + 11$, neither of which look hopeful.

$31 = 9 + 9 + 9 + 4$, a second way to do it.

$32 = 16 + 16$, but I could find no other way.

$33 = 16 + 16 + 1$ using 32 that one was easy!
$\quad = 25 + 4 + 4$ now I can use these for 34.

$34 = 16 + 16 + 1 + 1$ (I am omitting all the intermediate calculations.)

$\quad = 25 + 4 + 4 + 1$
$\quad = 25 + 9$
$\quad = 16 + 9 + 9$

$35 = 25 + 9 + 1$
$\quad = 16 + 9 + 9 + 1$

$$36 = 36$$
$$= 25 + 9 + 1 + 1$$
$$= 16 + 16 + 4$$
$$= 9 + 9 + 9 + 9$$

$$37 = 36 + 1$$
$$= 16 + 16 + 4 + 1$$

and so it goes on. I have done more than asked, having found different ways of representing some of the numbers. I noticed that it was easy to get caught in a rut and not to notice other ways of getting a number as the sum of four or fewer squares. My specializing has suggested that it might be true that *any* number is the sum of four or fewer squares, but it is certainly false that three squares would suffice, since 31 requires four squares.

1.2   Having tried some examples with whole numbers, fractions and decimals, a conjecture will have emerged. Now try it in general. Let one number be x, so that the other is $1 - x$. Squaring x and adding $1 - x$ gives $x^2 + 1 - x$. Squaring $1 - x$ and adding x gives the same result. Thus the two *are* always the same. Try generalizing to products, or to sums other than 1, or to three numbers. The specified computation may need altering as well.

1.3 $\sqrt{1} = 1$

$\sqrt{11} = 3.3166248$ from my calculator

$\sqrt{121} = 11$

$\sqrt{1221} = 34.9428$

$\sqrt{12321} = 111$

This seems to suggest a pattern. I conjecture that $1\,234\,321$ will have 1111 as its square root, and that the square root I was asked for is 111111111, because the number of 1s corresponds to the largest digit in the middle of the number. I find myself wondering what the square of ten 1s, $1\,111\,111\,111$, might look like....

1.4   The only thing to do is to specialize.
$$1 + 2 + 4 = 7 \text{ prime}$$
$$1 + 2 + 4 + 8 = 15 = 3 \times 5 \text{ composite}$$
$$1 + 2 + 4 + 8 + 16 = 31 \text{ prime}$$
$$1 + 2 + 4 + 8 + 16 + 32 = 63 = 3 \times 3 \times 7 \text{ composite}$$
$$1 + 2 + 4 + 8 + 16 + 32 + 64 = 127 \text{ prime}$$
$$1 + 2 + 4 + 8 + 16 + 32 + 64 + 128 = 255$$
$$= 3 \times 5 \times 17 \text{ composite}$$
$$1 + 2 + 4 + 8 + 16 + 32 + 64 + 128 + 256 = 511$$
$$= 7 \times 73 \text{ composite---oops!}$$

Perhaps Tartaglia did not specialize far enough! How did I know to keep going? I have enough experience with primes to know that there are very few simple statements about primes which are likely to be true. The conjecture needs modifying, but it is not at all clear how to modify it.

N	6N − 1	6N + 1
1	5 prime	7 prime
2	11 prime	13 prime
3	17 prime	19 prime
4	23 prime	25 = 5 × 5
5	29 prime	31 prime
6	35 = 5 × 7	37 prime
7	41 prime	43 prime
8	47 prime	49 = 7 × 7
9	53 prime	55 = 5 × 11
10	59 prime	61 prime
11	65 = 5 × 13	67 prime
12	71 prime	73 prime
13	77 = 7 × 11	79 prime
14	83 prime	85 = 5 × 17
15	89 prime	91 = 7 × 13
16	95 = 5 × 19	97 prime
17	101 prime	103 prime
18	107 prime	109 prime
19	113 prime	115 = 5 × 23
20	119 = 7 × 17	121 = 11 × 11

Oops—neither of those last two are prime. Two things to learn from these examples: *never* believe your conjecture until you have found a convincing justification or at least feeling for *why* it might be true. Simply accumulating facts is not enough. Keep clear what you KNOW and what you WANT. I several times wrote down that a number was prime, 119 in particular, because I got into the habit of looking at a number and wishing it to be prime. I had to double check each time to be sure.

1.5   I tried squares and rectangles first. Then I tried a triangle and realized that I couldn't do the computations! I began to draw figures with long spindly bits joining massive bits. My favourite is shown here, together with an example based on a circular ring with a piece missing, but there are many other examples which show that the centre of gravity has nothing to do with lines bisecting area, even though it is a reasonable first conjecture from symmetric figures.

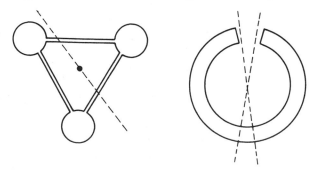

1.6   The true formula is

$$F = 32 + \tfrac{9}{5}C$$

whereas the formula being offered is

$$F = 30 + 2C$$

I experimented with various values for C, and found that for values between 5 and 15 degrees Centigrade, the two formulas give values that are no more than 1 degree Fahrenheit apart. The temperature in the UK is not confined to that range, but it is not a bad attempt.

A little bit of algebra (which you were not asked to do!) shows that

$$|(30 + 2C) - (32 + 9C/5)| < 1$$

means that

$$|C/5 - 2| < 1$$

or

$$-1 < C/5 - 2 < 1$$

or

$$1 < C/5 < 3$$

or

$$5 < C < 15$$

In other words, *only* for values between 5 and 15 degrees Centigrade will the announced formula give values within 1 degree Fahrenheit.

1.7   Specialize. If 2 divides the product of two numbers, say 14, then 2 certainly divides one of them, in this case because $14 = 2 \times 7 = 1 \times 14$. The same is true of 3 dividing a number, and more generally of any prime. The reason is that primes are indivisible, they cannot be broken up, so if a prime is to divide a number, it must divide one of the factors.

If I try a number that is not prime, like 4, then certainly 4 divides 2×2, but 4 does not divide either factor. The statement must be modified to refer to prime numbers dividing a product forcing the prime to divide a factor.

So far I have confined my attention to positive integers. What happens if rational or real numbers are admitted? It all depends on how I carry over the idea of 'divides'. With whole numbers it means divides exactly without remainder. If rational numbers are admitted, then 2 divides into 3 to give $\tfrac{3}{2}$, does if not? It seems best to leave the original statement referring to integers—though perhaps with some work the idea could be extended. You might like to consider what happens if you confine your attention to the set of numbers that have a remainder 1 when divided by 3. For example 4, 10 and 25 are in the set, and none is a product of members of the set other than itself with one, so 4, 10 and 25 are all 'prime'. However, $4 \times 25 = 10 \times 10$, so we find 4 divides 100 but 4 does not divide 10 or 10. You might like to investigate the 'primes' in this system, always confining yourself to numbers having remainder 1 when divided by 3.

1.8   Trying various values of x, I first tried positive values, and found that the statement seemed correct. Negative values of x were fine until I tried $x = -4$. It is easy to be confined by what you want to be true, and to forget or fail to see other possibilities.

1.9   I specialized by drawing diagrams, and concluded that twelve triangles are needed, because triangles have to be glued together along complete edges, and there must be a proper hole. Then it occurred to me to try three dimensions. I'm not really sure what an annulus is, but I cut out a strip of triangles, started folding them, and discovered...(you do it!).

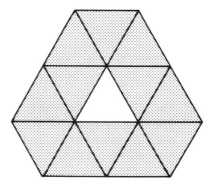

Section 2   Generalizing

Dinner

I want some way to compare and combine the rates of eating. The lion eats $\frac{1}{2}$ of a sheep per hour, the wolf $\frac{1}{3}$ and the dog $\frac{1}{5}$. Thus in T hours, they have eaten

   $T(\frac{1}{2} + \frac{1}{3} + \frac{1}{5})$ of a sheep

I want T when one sheep has been eaten, i.e. when

   $T(\frac{1}{2} + \frac{1}{3} + \frac{1}{5}) = 1$

or

$$\frac{1}{T} = \frac{1}{2} + \frac{1}{3} + \frac{1}{5}$$

I have carefully avoided doing arithmetical computations, because now I can generalize—replace 2 by L for the lion's time of eating one sheep, and so on.

Generalized Ratios

Instead of just adding numerators and denominators, it occurred to me to multiply by positive factors, to obtain things like

$$\frac{2A + 3C}{2B + 3D}, \text{ and more generally } \frac{xA + yC}{xB + yD}$$

where x and y are positive.

It also occurred to me to use more ratios like $\dfrac{E}{F}$ also equal to $\dfrac{A}{B}$, with positive factors:

$$\frac{xA + yC + zE}{xB + yD + zF}$$

I confine myself to positive factors, because otherwise I might get a zero in the denominator.

It also occurred to me to let go of the equality restriction, and look at inequalities. For example,

if $\dfrac{A}{B} < \dfrac{C}{D}$, then where do $\dfrac{A+C}{B+D}$ and $\dfrac{2A+3C}{2B+3D}$ fit in?

Exercise 4.5 investigates this further, but you might like to know that it lies at the heart of a good deal of Greek mathematics.

2.1(a) The first thing I did was to check the given statements. I seem to be faced with examples of numbers whose sum and product is the same, and one of the numbers is an integer. Having written down two more examples for myself, I generalized. Let A and B be two numbers of the sort I seek.

I know that $A + B = A \times B$

I also know that A is to be an integer, so I want to find the corresponding B. Solving for B gives me

$$B = \frac{A}{A - 1}$$

This formula specializes to the cases given, and enables me to generate other examples, not confined to integers.

(b) Similar work yields the second pattern, where the sum is the same as the product of one number and the square of the other. I can also see how to generate plenty of other similar examples.

2.2 The pattern seems clear enough—until I try to write something down! If there are k 6s in front of a single 7, then I suspect that the square will consist of $(k + 1)$ 4s followed by k 8s and a final 9.

I checked the three examples and one more before my calculator switched into exponential notation, and so stopped me checking the pattern.

It is possible to justify the conjecture using knowledge of how to sum a geometric progression, or by carefully writing out rows of the long multiplication. You were not asked to do either! Here is the series approach.

$$67 = 6 \times (10 + 1) + 1$$

$$667 = 6 \times (100 + 10 + 1) + 1$$

$$6667 = 6 \times (1000 + 100 + 10 + 1) + 1$$

The number with k 6s to the left of a single seven, is one more than 6 times the sum of the first $k + 1$ powers of 10, starting at $10^0 = 1$. I can compute this sum, using the formula

$$1 + r + r^2 + \ldots + r^t = \frac{r^{t+1} - 1}{r - 1}$$

for the sum of a geometric progression, as

$$\frac{10^{k+2} - 1}{10 - 1} = \frac{10^{k+2} - 1}{9}$$

I want to look at the square of

$$6 \times \frac{10^{k+2} - 1}{9} + 1 = 2 \times \frac{10^{k+2} - 1}{3} + 1$$

which I shall call S for short. Then I want to find the pattern of 4s, 8s and 9 in S^2:

$$\left(2 \times \frac{10^{k+2} - 1}{3} + 1 \right)^2$$

$$= 4 \times \frac{10^{2k+4} - 2 \times 10^{k+2} + 1}{9} + 4 \times \frac{10^{k+2} - 1}{3} + 1$$

Now I look at what I want it to be, in order to see where to go from here. I want to end up with similar sums of geometric progressions for the repeated 4s, and the repeated 8s, so I want to get both denominators in the form 9, which is $10 - 1$. Furthermore, I am looking for terms of the form $10^{t+1} - 1$ coming from sums of powers of 10, so after some time looking at the mess of stuff, I find

$$4 \times \frac{10^{2k+4} - 2 \times 10^{k+2} + 1}{9} + 4 \times \frac{10^{k+2} - 1}{3} + 1$$

$$= 4 \times \frac{10^{2k+4} - 2 \times 10^{k+2} + 1}{9} + 12 \times \frac{10^{k+2} - 1}{9} + 1$$

$$= 4 \times \frac{10^{2k+4} - 1}{9} + 4 \times \frac{10^{k+2} - 1}{9} + 1$$

which is another way of saying $(2k + 3)$ 4s added to $(k + 1)$ 4s added to 1. Adding them together gives $(k + 1)$4s followed by k 8s followed by 9.

2.3 This is very similar to the preceding exercise. In fact, the numbers being squared in 2.4 are one less than double the numbers being squared in 2.5.

2.4 First I check to see if the arithmetic is correct, and in the process of writing down the sums, begin to see patterns. I find it useful to write down the features which strike me—that the numbers are consecutive, that there is one more on the right-hand side than on the left of each one, and that there is one more term on each side than in the preceding line. This is enough to tell me how to write down the next few cases. Then I notice that the first term is a square... in fact, the square of the number of the line it is on. I am bold, and conjecture that the Nth line will begin with N^2. There will be N more terms on the left, and N terms on the right. (Specialize to check!) The result is a conjecture that

$$N^2 + (N^2 + 1) + (N^2 + 2) + \ldots + (N^2 + N)$$
$$= (N^2 + N + 1) + (N^2 + N + 2) + \ldots + (N^2 + N + N)$$

I notice that the next consecutive integer is $N^2 + 2N + 1$, which is a perfect square, ready to start the next line, so I feel fairly confident that I have the pattern right at least. I still do not know whether the two expressions are always equal!

You were asked only to generalize the pattern, but I notice that if I compare the two expressions, the last terms differ by N, as do the next to last,... down to the second term of the first expression and the first term of the second expression. There are N terms in the second expression, so the second expression is N^2 more than the corresponding terms in the first, which has an extra N^2 to balance. The two expressions *are* always equal.

2.5   This one is similar in flavour to Exercise 2.6. First, I check the arithmetic, and at the same time look for patterns. I knew the first line, was a bit surprised by the second, and frankly dubious about the third, but they all checked. The features I noticed were that the left-hand sides have one more term than the right-hand sides, and that each line has one more term on each side than the preceding line. It is not so easy to see how to start each line. Let me call the starting number S. Then I suspect that the Rth row or line looks like:

$$S^2 + (S + 1)^2 + (S + 2)^2 + \ldots + (S + R)^2$$
$$= (S + R + 1)^2 + (S + R + 2)^2 + \ldots + (S + R + R)^2$$

I checked that this does specialize back to the cases I was given for $R = 1, 2$ and 3—so now all I need are the starting numbers.

The first terms of the lines I am given are 3, 10, 21. These are rising by 7, then 11. A very bold guess might be to try a further rise of 15 on the grounds that 11 is 4 more than 7, but it seems pretty far-fetched. It also works!

$$36^2 + 37^2 + 38^2 + 39^2 + 40^2 = 41^2 + 42^2 + 43^2 + 44^2$$

How can I get hold of 3, 10, 21, 36? Try relating them to the number of the line that they start:

row	1	2	3	4
starts	3	10	21	36

I notice that the row number divides the starting number:

$$3 = 1 \times 3, \ 10 = 2 \times 5, \ 21 = 3 \times 7, \ 36 = 4 \times 9$$

Staring at those factors for a while, and wondering how to deduce the second factor, I noticed that

$$10 = 2 \times 5 = 2 \times (2 \times 2 + 1)$$
$$21 = 3 \times 7 = 3 \times (2 \times 3 + 1)$$
$$36 = 4 \times 9 = 4 \times (2 \times 4 + 1)$$

I am now ready to conjecture that the Rth row starts with $R(2R + 1)$. The Rth row should then read

$$S^2 + (S+1)^2 + (S+2)^2 + \ldots + (S+R)^2$$
$$= (S+R+1)^2 + (S+R+2)^2 + \ldots + (S+R+R)^2$$

where $S = R(2R+1)$. I checked to see that this does specialize back to the initial cases. I have got only a conjecture here, because I have not shown that the two expressions are always the same.

2.6 Having checked the arithmetic, I tried the same procedure on another pair of ratios and found the same result. With my experience with *Ratios* in mind. I generalized boldly!

Suppose $\dfrac{A}{B} = \dfrac{C}{D}$

I conjecture that

$$\frac{pA + qB}{rA + sB} = \frac{pC + qD}{rC + sD}$$

where p, q, r and s can be positive, negative or zero as long as the denominators are not zero.

Putting $\dfrac{A}{B} = \dfrac{C}{D} = k$ and calculating as before proves that the conjecture is indeed correct.

2.7 The purpose of this question is simply to raise questions, not to begin answering them. Here are a few suggestions based on stressing-in-order-to-question the significance of individual words in each of the statements.

Emphasis on *two* suggests taking three or more numbers. Emphasis on *odd* suggests trying even numbers, or perhaps numbers like $5N + 1$. Emphasis on product suggests other computations, perhaps along the line suggested in Still More Remainders.

In place of the sum, try the product, or more complicated expressions; try more than two numbers; try to find an expression which is minimized rather than maximized when the numbers are equal.

Try interchanging perimeter and area; try making the area bigger rather than smaller; look for a connection with other similar statements, such as the previous one; try triangles or other figures in place of rectangles; try three-dimensional figures.

Try products instead of sums; try moving to quadrilaterals or other figures; try three dimensions; try areas in place of lengths—for example the area of one face of a tetrahedron is presumably smaller than the sum of the areas of the other three faces....

Variations and extensions of Pythagoras' theorem are described in detail in the next section.

Try other calculations—for example Ptolemy claimed that

$$|AB| \times |CD| + |AD| \times |BC| > |AC| \times |BD|$$

with equality only when A, B, C, D lie in order on a straight line or on a circle! Try describing the circle; try looking at lengths rather than angles; try other polygons; try three dimensions.

Section 3   Specializing and Generalizing Together

Recovering 8, 15, 17 from the General Formula for Pythagorean Triples

Since $2MN$ is even, it probably corresponds to 8, thus $MN = 4$. Try $M = 2$ and $N = 2$. No that was silly, because the first term of the triple would be zero. Try $M = 4$ and $N = 1$. The corresponding triple is 15, 8, 17 as requested.

Constant Perimeter Tetrahedra

First, I must be clear about the terms. A tetrahedron is a triangular-based pyramid. It has four triangular faces and six edges. Having drawn myself a picture, I could not see any way of using numbers effectively. The only thing I could see to do was to label the edges and write down what I knew.

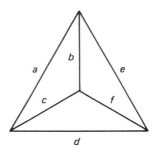

Each edge belongs to two faces, so there is a fair amount of information floating around. Let P represent the perimeter of each of the faces. Then

$$P = a + b + c$$

$$P = a + d + e$$

$$P = c + f + d$$

$$P = b + f + e$$

I WANT to see if there are any restrictions on the tetrahedron, and the particular perimeter is irrelevant, so I want to eliminate P. Before embarking, there must be some sort of pattern in the equations—for example, each letter occurs twice overall. Why not add them all together!

$$4P = 2(a + b + c + d + e + f)$$

I KNOW that $a + b + c = P$, so I can simplify and substitute in the last equation and get

$$2P = P + d + e + f$$

or

$$P = d + e + f$$

Now that is a new one—how does it compare with the original four? It overlaps quite nicely with three of them!

$$P = d + e + f$$
$$P = a + d + e$$
$$P = c + f + d$$

$$P = b + f + e$$

These immediately tell me that $a = f$, $c = e$ and $b = d$. Looking at the tetrahedron shows me that opposite edges of the tetrahedron must be equal... and more! The four triangles all have the same three edge-lengths, so all four faces are congruent.

Immediately, I must ask myself if such tetrahedra exist—surely I can make a tetrahedron from four copies of a triangle? Exercise 3.1 pursues this.

3.1 To make a tetrahedron, it seems awfully tedious to cut out four triangles and start glueing them together—I should be able to draw the triangles so as to be able to cut out a shape and fold the triangles into the tetrahedron. After some experimentation, I found that I was really starting with a triangle and dividing it into four congruent triangles.

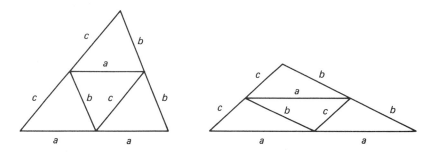

The first few examples seemed to fold up without difficulty. Then I tried an obtuse-angled triangle, and it did not want to work. The flaps folded right over flat and still did not touch. I tried an approximately right-angled triangle, and the flaps just about touched. I am led to conjecture that the tetrahedron can be made only if the face triangles are acute-angled. I leave investigation of this conjecture until Exercise 4.2.

3.2 The four equations I am given involve the sums of three squares equal to a square, which can be expressed as

$$P^2 + Q^2 + R^2 = S^2$$

I want to be able to find integer solutions. The only thing I know that seems relevant is the Pythagorean triples. If Q, R and S were one of those triples, then I could try to make either Q or R the third member of another Pythagorean triple, and glue them all together.

I was unable to extend $3^2 + 4^2 = 5^2$, but I used it to extend

$$5^2 + 12^2 = 13^2 \text{ to } 3^2 + 4^2 + 12^2 = 13^2$$

It seems a cumbersome way to proceed.

I have been asked to do some algebra on X, Y and Z. By analogy with the Pythagorean triples, the two terms $(2XY)^2$ and $(2XZ)^2$ are put in to convert the corresponding negative terms in the expanded square of the first expression, into the positive terms needed for the right-hand side. It is a generalization of the pattern for Pythagorean triples, and could even be extended, presumably, to more terms.

I specialized the formulas to produce the examples given (for example,

Specializing

taking $X = Y = Z = 1$ gives $1^2 + 2^2 + 2^2 = 3^2$), so the formulas do constitute a generalization! The last part claims that there are examples of three squares adding to a square which cannot be generated by the formula. Let us investigate.

I want to see why 3, 4, 12, 13 cannot be produced by the given formula. The formula has two even terms, so they must correspond to the 4 and the 12. Does it matter which one is which? Surely not, because Y and Z are interchangeable in the formula.

Put $4 = 2XY$ and $12 = 2XZ$

Since X, Y, Z are supposed to be integers (and they might as well be positive), the only possibilities are that

$X = 2$ so $Y = 1$, $Z = 3$, giving $(4 + 1 - 9)^2 + 4^2 + 12^2 = (4 + 1 + 9)^2$

or

$X = 1$ so $Y = 2$, $Z = 6$ giving $(1 + 4 - 36)^2 + 4^2 + 12^2 = (1 + 4 + 36)^2$

neither of which is the expression being sought. I am left wondering if there *is* a general quadruple which will generate *all* possible integer quadruples, as an analogue with the Pythagorean triples. One student noticed that

in $2^2 + 3^2 + 6^2 = 7^2$, $3 - 2 = 1 = 7 - 6$ and $2 \times 3 = 6$

in $3^2 + 4^2 + 12^2 = 13^2$, $4 - 3 = 1 = 13 - 12$ and $3 \times 4 = 12$

and investigated generalizations of this pattern.

3.3   Start by drawing the shapes yourself. It is amazing how this reveals patterns that may not be obvious when you look at someone elses pictures. Try to express a method for drawing the next, and the next, in general. Specialize, by working out the numbers of dots in the first few patterns of a sequence, generalize, and then use that to predict the number of dots and for the 37th shape. Even if the numbers seem to follow a pattern, you still have to verify that the number pattern reflects or captures what is happening in the diagrams. In the first sequence, it can help if you think about completing each triangle to a rectangle using a second copy. In the second sequence, the numbers appear to be squares, so try to find a rearrangement that makes a square, and which applies to the general case. The shapes in the third sequence can be seen as four arms. The shapes in the fourth sequence can be seen as being made up of several copies of the shapes in the first sequence, or as a sequence of layers added on, or in many other ways.

Try making up your own sequences, and finding the number of dots needed for the general term. Try taking any sequence, and putting a border of dots all the way round each shape in turn. Then predict the number of dots needed for the border in general.

3.4   **Family Tree**

I began by drawing myself some simple family trees and computing 'parts'. If my father is Cherokee and my mother Algonquin, then I am half Cherokee and half Algonquin. That much seems clear. Perhaps if my parents are part this and part that..., presumably my 'parts' are found by averaging corresponding parts of my parents. After several specific calculations I realized that the only denominators I can possibly get are powers of 2. Unless Gilgamesh had three parents, or infinitely many parents, I cannot see how the $\frac{2}{3}$ and $\frac{1}{3}$ can arise.

3.5   Crossless

The first thing to note is that the question is ambiguous. Are the points already given, and must the maximum be the maximum in the worst possible case for me, or am I permitted to place the points so as to maximize the number of non-crossing segments? The latter seems to give more freedom, so I looked at it.

I want to arrange P points in the plane so that as many pairs as possible can be joined by straight-line segments, on two of which cross. Let L be the number of non-crossing lines. Specialize.

$P = 2$ $L = 1$ no choice

$P = 3$ $L = 3$ no choice. Too early to conjecture

$P = 4$ $L = 5$

some choice, but expected $L = 6$

AHA!

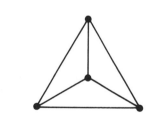

$L = 6$ by exploiting the choice

$P = 5$ $L = 7$ No! $L = 8$

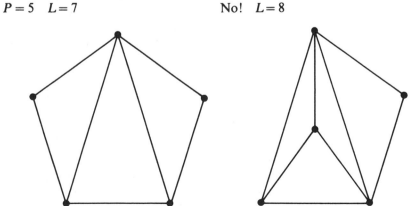

$P = 6$ $L = 11$, glue two copies of $P = 4$ along an edge

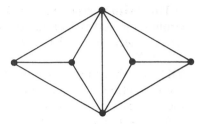

Looking at 1, 3, 6, 8, 11, I see that it alternates differences of 2 and 3. Why is that? $P = 4$ was the first real choice, and all possible lines are present. Try $P = 5$ again.

Aha! Put the fifth point *inside* one of the old triangles: $P = 5$, $L = 9$. Now I see a general pattern and a general strategy to support it.

Put each new point inside an old triangle and draw three new lines. For $P = 6$, $L = 12$ is possible this way. I predict that P points will produce... the first two points are special, but after that there are three lines per point, so $1 + 3(P - 2) = L$ is my first (but wrong) conjecture. The initial 1 copes with the single line when $P = 2$. I must stipulate that my conjecture can only work for $P > 1$. Check it by specializing—it's wrong—so try again. For $P > 3$, I get 3 lines for each new point, and there are 3 lines for $P = 3$, so try $3 + 3(P - 3) = L$. At least this specializes correctly as long as $P > 2$!

Is it always going to work? My strategy can certainly yield that many non-crossing lines, but is that the very best possible? Might not some other approach yield more?

Section 4   Convincing Yourself and Others

4.1   Since $12 = 2 \times 2 \times 3$, according to the theory developed, the numbers with 12 divisors have the form

$$p^{2-1}q^{2-1}r^{3-1}, \quad p^{4-1}q^{3-1} \quad \text{or} \quad p^{12-1}$$

where p, q and r are distinct prime numbers. I WANT the smallest possible product, so I choose the smallest primes to go with the largest indices in each case. This leads me to

$$5 \times 3 \times 2^2, \ 2^3 \times 3^2 \text{ and } 2^{11}, \text{ or } 60, 72 \text{ and } 2048$$

Thus 60 is the smallest number with 12 divisors. CHECK!

The divisors of 60 are 1, 2, 3, 4, 5, 6, 10, 12, 15, 20, 30 and 60. When I first wrote them down, I forgot 20, and it took me a long time to find it—I kept looking because I believed my theory more strongly than my arithmetic!

4.2 I KNOW that angles are important, because the conjecture mentions acute and obtuse. I also KNOW from my specializing with paper triangles that sometimes the flaps meet, and sometimes they do not. I WANT to see when a tetrahedron is formed, so look at two flaps that must meet.

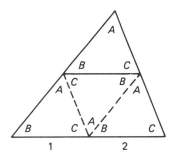

The angles B and C must together be large enough so that when the flaps are folded along the dashed edges, the edges marked 1 and 2 must actually meet in space. Thus B and C must add to more than the third angle at that point, which is A. I also KNOW that the sum of the angles of any triangle is 180°, and since two add to more than the third, the third must be less than 90°. Put algebraically,

$A < B + C$ for the flaps to meet properly,

$A + B + C = 180,$ angle sum of any triangle,

thus

$A + A < 180, \quad \text{so } A < 90$

The same thing happens between the other pairs of flaps, so all angles must be less than 90°. In other words, the triangle must be acute.

4.3 I KNOW that $\sqrt{2}$ is irrational.

I KNOW that $\sqrt{2}^{\sqrt{2}}$ is an irrational raised to an irrational power. Is that all there is to it? I WANT an irrational raised to an irrational to be rational. What about $\sqrt{2}^{\sqrt{2}}$? I don't know whether it is rational or irrational, though I suspect it is irrational. I have been told to look at the sequence, so let me continue my rather elementary specializing.

I KNOW that

$$(\sqrt{2}^{\sqrt{2}})^{\sqrt{2}} = \sqrt{2}^{\sqrt{2} \times \sqrt{2}}$$
$$= \sqrt{2}^{2}$$
$$= 2$$

This is certainly rational, but a bit of a surprise. I WANT an irrational raised to an irrational to be rational. Either $\sqrt{2}^{\sqrt{2}}$ is rational, in which case I have what I WANT, or it is irrational, in which case itself raised to $\sqrt{2}$ (which is irrational) is rational, which is what I WANT. In either case I get what I WANT, so the claim has been justified. Notice that I don't actually know *which* example is the one I want, but I do know that it must be one of them! To generalize, try the same idea with cube roots, etc.

4.4 I KNOW that a rational number is a terminating or repeating decimal. I WANT a terminating decimal between two given rationals. Specialize by starting with terminating decimals:

> try 0.1234   and   0.1 567

I can choose anything in between, say 0.12348, indeed anything finite on the end of the smaller number (generalizing). Can anything go wrong? Specialize again, trying to force my idea to go wrong:

> try 0.12345   and   0.12346

Not much room there—but I can still stick anything finite on the back of the smaller number. Hmm.

> Try 0.1234   and   0.12340000001

Reading the decimal places from left to right, eventually one of the digits of the larger number must be different from and bigger than the corresponding digit of the smaller number, and I can squeeze in my number with a digit in between. That's fine, but what about the repeating decimals—I had forgotten about them. It is still true that at some point the digits of the two numbers must disagree for the first time, and that is where I can prise them apart with a terminating decimal, which is of course rational. Later, someone showed me

> 1.49999999...   and   1.5

and asked me about inserting a number in between. My reply was that there *is* no number in between, that they are two representations of the same number, just as 2/3 and 4/6 are two different representations of the same number.

Now, given two irrational numbers, I KNOW *only* that they have non-terminating, non-repeating decimal representations. I can still squeeze a number in between the way I did before—that argument made *no use* of the fact that the two given numbers were rational. I actually showed (generalizing) that between *any* two numbers, there is a rational number (indeed, a terminating decimal).

Given two irrational numbers, I WANT to squeeze an irrational in between—but what does an irrational look like? I KNOW only that it is non-terminating and non-repeating. I KNOW I can get a rational in between any two numbers, so perhaps I can change it into an irrational by giving it a tail. What sort of tail could I give it that was guaranteed not to repeat or terminate? I started writing down a sequence of digits:

> 101001000100001000001...

Each new 1 is followed by a longer sequence of 0s than has previously occurred, so there is no fear of repeating, and it goes on forever, so it doesn't terminate. All I have to do is stick that on the end of the rational which lies between the two given numbers, and I have what I WANTED.

I have actually shown that between any two numbers there is at least one rational number, and at least one irrational.

4.5   Specialize first!

$$\frac{1}{2} < \frac{3}{4}, \quad \text{so} \quad \frac{1+3}{2+4} = \frac{4}{6} = \frac{2}{3} \quad \text{and} \quad \frac{1}{2} < \frac{2}{3} < \frac{3}{4}$$

Also,

$$\frac{2 \times 1 + 3 \times 3}{2 \times 2 + 3 \times 4} = \frac{11}{16} \quad \text{and} \quad \frac{1}{2} < \frac{11}{16} < \frac{3}{4}$$

The best way to deal with the generality is to do general calculations. To show that one number is bigger than another, it is usually easier to subtract the smaller one from the larger. Following this on the more complicated ratio:

I KNOW

$$\frac{2A + 3C}{2B + 3D} - \frac{A}{B} = \frac{2AB + 3BC - 2AB - 3AD}{B(2B + 3D)}$$

$$= \frac{3(BC - AD)}{B(2B + 3D)} \quad \text{I WANT this positive}$$

I KNOW that $\frac{A}{B} < \frac{C}{D}$, that is $\frac{C}{D} - \frac{A}{B} > 0$ or $\frac{BC - AD}{BD} > 0.$

So... I WANT a $(2B + 3D)$ on the bottom, and I have a D instead. Suppose all the numbers are positive—it will surely work then, because the denominators are positive forcing $BC - AD$ to be positive, so the difference I began with will be positive. It all hinges on $BC - AD$.

Now I want to see if I can defeat the inequality using negatives. Try

$$\frac{3}{-4} < \frac{1}{-2}, \text{ which produces } \frac{2 \times 3 + 4 \times 1}{2 \times (-4) + 3 \times (-2)} = \frac{9}{-14}$$

and $\dfrac{3}{-4} < \dfrac{9}{-14} < \dfrac{11}{-2}$, which is fine.

I WANT a minus sign on top as well, say in C.

$$\frac{3}{-4} < \frac{-1}{2} \text{ produces } \frac{2 \times 3 + 3 \times (-1)}{2 \times (-4) + 3 \times 2} = \frac{3}{-2}$$

and defeats the conjecture.

This shows that the original assertion has to be modified. I shall be content with assuming that A, B, C and D are all positive. There might easily be a refinement that permits negatives, but I shall leave it there.

Looking back over my work, I discovered that I entirely forgot about the other half of the inequality. When I worked it through, there were no more surprises, fortunately!

Trying to place it in a more general context, I replaced the 2 and the 3 in the original question, with X and Y, both assumed positive. I expected to have to place another condition on X and Y, but the algebra showed it to be unnecessary.

A journey consisting of two parts covering a distance of A in time B and then a distance C in time D would result in an average speed of A/B for the first part, and an average speed of C/D for the second part. The average speed overall must lie between the average speeds for the two parts—a re-statement of the conjecture. Inserting X and Y just magnifies the distances and time in each part, but still the average overall lies between the averages for each part.

The parallelogram $(0, 0)$, (XB, XA), (YD, YC), $(XB + YD, XA + YC)$ must have the slope of the diagonal from $(0, 0)$, between the slopes of the edges from $(0, 0)$, which is a re-statement of the conjecture.

Section 5   When is an Argument Valid?

One is the Largest Whole Number

In the case of Steiner, it can with difficulty be shown that there is a largest area, and the problem is to show that this largest area is in fact a segment of a circle. In the largest number conjecture, the argument breaks down when it is claimed that you must eventually get close to one. This claim already assumes that the original conjecture is true, namely that there is a largest number.

5.1   The function as specified has no domain, because for no real value of x are $1 - x^2$ and $x^2 - 4$ both positive. Since there can be no x for which the statement is false, it is in fact true that the function never exceeds 2 (or any other value!).

5.2   While it is certainly true that log is an increasing function, it pays to be clear about the domain of a function before beginning work on it. In this case, $\sin(x)$ always lies in the interval $[-1, 1]$, and log can accept only positive values for its domain. Since the log of a number less than or equal to 1 is at most 0, the second log function cannot accept any of the images of $\log(\sin(x))$. Thus the composite function has no domain at all. In particular, it has no maximum value (indeed no value at all!).

5.3   The argument offered is somewhat succinct. I WANT to know what x and y must be. I KNOW that they satisfy the equation. I WANT two squares to add to 0. Since squared numbers are never negative, they must both be 0 in order to add to 0. Therefore $x = 2$ and $y = -3$.

5.4   The argument offered is argument by specializing. In this case, *all* possible cases have been considered: since all arithmetic is being done modulo 3, any number is congruent to 0, 1 or 2 modulo 3. The argument is valid as it stands, but compare it with Exercise 5.11, which tries to argue the same way!

5.5   A great deal of care is needed when writing down equations from words. It is tempting to catch sight of '14 times...students', but in fact the words omitted, 'as many', also have meaning. It may help to translate the words into more careful English first, paying particular attention to what it is that S and L stand for.

The number of students is 14 times the number of lecturers.

Now it is much easier to translate into symbols because all the English words have mathematical versions, whereas there is no mathematical version of 'as many as'.

5.6 This is an example of 'argument by calculator'. The number is an integer only to the accuracy of the calculator. It is not, in fact, an integer.

5.7 The argument offered needs expanding with KNOW and WANT, but it shows that the edges of the octagon are all equal in length. The claim is that the octagon is regular—what does regular mean? It means equal sides *and* equal angles. Further computation is required! Unfortunately, the angles turn out to be slightly different. The symmetry shows that the angles at A, G, E and C are equal, as the angles at H, F, D, B, but the two values are distinct.

5.8 The proposed argument starts with what it WANTS and works away, modifying the WANTS, until it reaches what is KNOWN. A convincing argument may indeed involve modifying what is WANTED, but in the final presentation of the argument, it must be clear that, starting with what is KNOWN, it is possible to reach what is WANTED. In this case, it is possible to add words to the steps to make the argument convincing by labelling clearly with WANT and KNOW. An alternative presentation, which is easier to follow, is obtained by reversing the steps, and again labelling clearly what is KNOWN at each stage. Reversing the steps does not always work though. Try following through the revised argument with the above steps reversed, but A, B, X, Y negative.

5.9 Look first at the case when $K = 2$. The argument as given seems to do little more than restate what is WANTED. Try 3 and 4 whose sum is 7. One of them is divisible by $K + 1 = 3$. What about 4 and 5. Aha, their sum is 9, which is divisible by 3. More generally, I WANT two consecutive numbers, so N and $N + 1$ seem reasonable choices. Now I WANT their sum, which is $2N + 1$. So far so good. Now I WANT to show that exactly one of these numbers is divisible by $K + 1$, which is 3. I detect three cases:

$$N \equiv 0 \bmod 3; \quad N \equiv 1 \bmod 3; \quad N \equiv 2 \bmod 3.$$

In each case, I compute $N + 1$ and $2N + 1$ to find that exactly one of N, $N + 1$, $2N + 1$ is divisible by 3. That much of the argument is reasonable.

The proposed argument goes on to claim that a similar argument works for other K. It seems a little early to generalize, since I have only one case to go by, so I specialize further.

Try $K = 3$. I choose 1, 2, 3, which seem like good candidates for three consecutive numbers, and their sum is 6. None of them is divisible by $K + 1$, which is 4. So much for that argument. I am led to wonder if it does work for any other value of K, but that is another question!

5.10 The result is certainly wrong—no, let me be careful—the result involves a curious meaning of 'no difference between'. This example demonstrates the necessity of having technical terms which are clear, and which correspond to intuition. There is an underlying assumption that if 'no difference' is added to 'no difference', then the result will still be 'no difference', which seems implausible!

5.11   There is nothing wrong with the specializations offered, but one must always be careful about generalizations. They make good conjectures, but they need supporting arguments before they are convincing. In this case, further specialization (two more cases, not just one!) shows that the pattern is *not* powers of two; it just looks like it at first.

5.12   The argument here is perfectly sound, but it relies without comment on the observation that the *number* of prime factors of an integer is a meaningful idea – so that it is not possible to get two different values depending on how you set about finding prime factors. See the resolution of Exercise 1.7.

5.13   The argument offered, where succinct, can be expanded by insertion of what is KNOWN and WANTED, to make it clearer. The generalization gratuitously thrown in at the end seems overly ambitious! Having written out the argument in more detail, it does extend to *A*, *B*, *C* and *D* all rational numbers but no further, because a crucial step in the argument is that $\sqrt{2}$ is not rational. (In fact, the argument does extend a bit further but it is too complicated to pursue here.)

5.14   It is certainly true that $2N$ represents any even number. However, in order to show that the sum of *any* two even numbers is even, I have to find a way to represent *any* two even numbers. The argument proposed adds an arbitrary even number to itself. To correct the argument, I WANT a representation of two perfectly arbitrary even numbers. One of them can be $2N$ and the other could be $2M$. Now the argument works. Note that as a special case of the argument, *M* and *N* might have the same value, but the presence of the *N* and *M* means that they might also have different values.

5.15   This is another example of the same error as in Example 5.12, but this time the assertion is false. For example $1 + 2 = 3$, which is not even.

5.16   My calculator gives $\cos(\sin(1)) = 1$ only when it is in degree mode. Working in degrees, it is not clear what to make of $\sin(1)$, because $\sin(1)$ is a number and not a number of degrees. Does composition of cos and sin really make sense when the calculator is in degree mode?

5.17   The argument offered purports to prove that two numbers are equal. The bridge between the two numbers comes from a diagram. To get π, I am told to look at the length of the semi-circular arcs, and at each stage it is hard to dispute that the total remains constant at value π. To get 2, I am told to look at the length of the diameter of the first semi-circle, and to agree that the arcs appear to be getting closer and closer to the diameter.

My faith in the argument rests on my sense of 'gets closer and closer'.

At each stage, the area encompassed between the circular arcs and the diameter is $\pi, 2\pi/4, 4\pi/16, \ldots$, which certainly gets smaller and smaller. However, look at the maximum height of each arc above the diameter. The first arc is one unit above the diameter. The second pair are each $\frac{1}{2}$ unit above, and so it continues. At every stage, a full unit can be broken into pieces which can then be placed upright under the arcs of that stage which seems to go against intuitive feelings for what 'getting closer and closer' actually means. Examples like this acted as a spur to mathematicians to clarify what we mean by two curves getting closer and closer to each other. The result is a batch of definitions and theorems forming a whole chapter of mathematics.

5.18 The argument begins by KNOWING that $A^2 + B^2 = C^2$, and deduces relationships which must necessarily hold. The only doubtful bit is when Z is made equal to 2. If X and Y are both odd or both even, then

$$C = Z(X^2 + Y^2)/2$$

$$B = Z(X^2 - Y^2)/2$$

$$A = ZXY$$

are all integers already. Putting $Z = 1$ will give me a triple which might not arise by putting $Z = 2$ and choosing X and Y differently. For example, with $X = 7$ and $Y = 3$,

$$C = (X^2 + Y^2)/2 = 29$$

$$B = (X^2 - Y^2)/2 = 20$$

$$A = XY = 21$$

However, it turns out that 20, 21, 29 *can* be produced by the triple

$$X^2 - Y^2, \quad 2XY, \quad X^2 + Y^2$$

by putting $2XY = 20$ and choosing $X = 5$ and $Y = 2$.

In fact, by going back to an early stage of the argument, the hiatus can be repaired. If $A^2 + B^2 = C^2$, and all common divisors have been removed, one of A or B must be even, but not both. (If neither, C^2 is divisible by 2 but not 4; if both, then C is also even.) Let A be the even one.

Now proceed as before. Since B and C are odd, $C + B$ and $C - B$ are both even. Let $A = 2a$.

Then $A^2 = 4a^2 = (C + B)(C - B)$

$$a^2 = \frac{C + B}{2} \times \frac{C - B}{2}$$

Then $\dfrac{C + B}{2} = X^2 Z$

$$\frac{C - B}{2} = Y^2 Z$$

and $\quad a^2 = X^2 Y^2 Z^2$

Since A, B and C have no common factors, Z must be 1.

Then $\quad C = C^2 + Y^2$

$\qquad B = X^2 - Y^2$

$\qquad a = XY$

$\qquad A = 2XY$

as before. Now I am sure that *all* Pythagorean triples are generated by

$$X^2 - Y^2, \; 2XY, \; X^2 + Y^2$$

You might like to investigate $X^2 + Y^2 + Z^2 = W^2$ as in Exercise 3.2.

For Your Amusement!

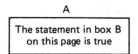

A

The statement in box B
on this page is true

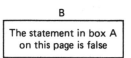

B

The statement in box A
on this page is false

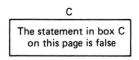

C

The statement in box C
on this page is false

D

There are two misstakes
in the statement in box D

FURTHER FOOD FOR THOUGHT

Expressing Generality

Sequences

For each of the following sequences, write down in words a rule for generating the configurations shown so that the sequence continues indefinitely. Then write down expressions for

> the number of objects (squares, triangles) needed in the nth configuration;

> the number of edges used;

> the number of exterior edges;

> the number of vertices;

> the number of exterior vertices.

There may well be connections between these that hold for every sequence! Make sure your expressions account for all the configurations shown in the sequence, as well as for your rule of drawing further ones! Your first expressions might be in terms of the numbers in preceding configurations, but try for direct formulae where possible.

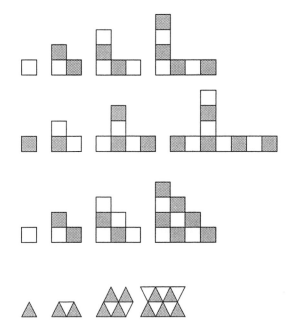

What is the same, and what different, about your rules for extending the last two sequences? Use that to generate other similar rules.

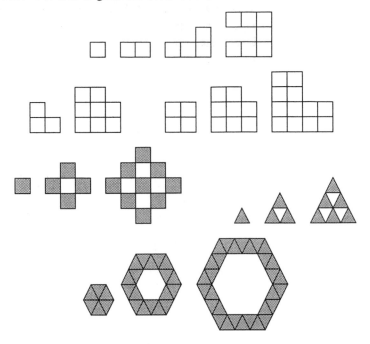

The following sequence of sequences can be worked on horizontally and vertically.

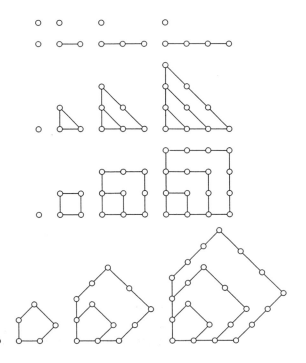

Borders

Take any sequence of shapes made from squares (or triangles) and place a border layer made up of squares (triangles) around each configuration. How many border elements are required? What if the borders are several layers thick? The process is illustrated here on two of the earlier sequences.

What features of a configuration do you need to know in order to work out how many border elements it requires?

Harder

These are based on a rule about adding new elements only if they do not touch another new one along an edge (at a vertex).

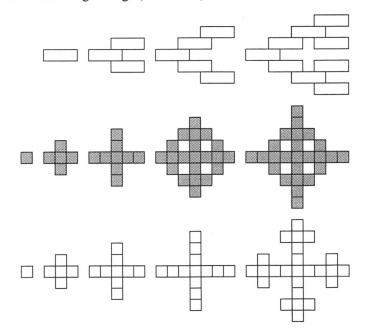

It is much more interesting to make up your own rule; it is very challenging to start with a sequence of numbers given by a rule and then to seek a 'matchstick' arrangement for which it is the expression.

Number Patterns

Blaise Pascal (1623-1662) constructed what is now known as Pascal's Triangle (shown below on the left), which appeared in the *Zhoubi suanjing* at least 2000 years ago in China, and is known there as the *GouGu* theorem, from *Gu* meaning *upright*, *Gou* meaning its shadow on the ground, and *Xian* the hypotenuse; thus $Gou^2 + Gu^2 = Xian^2$. Gottfried Wilhelm von Leibniz (1646-1716) constructed what is now known as Leibniz's triangle (shown on the right). Find at least two ways to generate successive rows of the triangles, which continue indefinitely:

$$1 \qquad\qquad\qquad \frac{1}{1}$$

$$1 \quad 1 \qquad\qquad \frac{1}{2} \quad \frac{1}{2}$$

$$1 \quad 2 \quad 1 \qquad \frac{1}{3} \quad \frac{1}{6} \quad \frac{1}{3}$$

$$1 \quad 3 \quad 3 \quad 1 \qquad \frac{1}{4} \quad \frac{1}{12} \quad \frac{1}{12} \quad \frac{1}{4}$$

$$1 \quad 4 \quad 6 \quad 4 \quad 1 \qquad \frac{1}{5} \quad \frac{1}{20} \quad \frac{1}{30} \quad \frac{1}{20} \quad \frac{1}{5}$$

Find connections between the two triangles.

Use Pascal's Triangle to conjecture and justify the sum of a finite number of terms from a diagonal, such as $1 + 3 + 6 + 10 \ldots + 55$.

Use Leibniz's triangle to conjecture and justify the sums of (infinite) diagonals, such as

$$\frac{1}{2} + \frac{1}{6} + \frac{1}{12} + \frac{1}{20} + \cdots .$$

Fraction Table

Find a way to generate successive rows of the following table, and a way to express the terms in the *n*th row:

$\frac{0}{1}$																$\frac{1}{1}$
$\frac{0}{1}$								$\frac{1}{2}$								$\frac{1}{1}$
$\frac{0}{1}$				$\frac{1}{3}$				$\frac{1}{2}$				$\frac{2}{3}$				$\frac{1}{1}$
$\frac{0}{1}$		$\frac{1}{4}$		$\frac{1}{3}$		$\frac{2}{5}$		$\frac{1}{2}$		$\frac{3}{5}$		$\frac{2}{3}$		$\frac{3}{4}$		$\frac{1}{1}$
$\frac{0}{1}$	$\frac{1}{5}$	$\frac{1}{4}$	$\frac{2}{7}$	$\frac{1}{3}$	$\frac{3}{8}$	$\frac{2}{5}$	$\frac{3}{7}$	$\frac{1}{2}$	$\frac{4}{7}$	$\frac{3}{5}$	$\frac{5}{8}$	$\frac{2}{3}$	$\frac{5}{7}$	$\frac{3}{4}$	$\frac{4}{5}$	$\frac{1}{1}$

Find and express other patterns and relationships.

Manifesting Particularity

Specialize the following conjectures; look for counter examples; then adjust them (if necessary) so they are correct. Convince yourself, a friend, a sceptic.

- Every straight line in the plane intersects the curve $y = x^t$ for any choice of real number t.
- There is a region in the plane which is intersected by every straight line, and such that that intersection is a single line segment.
- Any quadrilateral in the plane tessellates the plane
- Among any odd number of whole numbers the sum of all but some one of them is even.
- For any integer x there exists a positive integer y such that $x > y^2$; For any real number x there exists a positive real number y such that $x > y^2$
- Between any two rational numbers there lies a real number; between any two real numbers there lies a rational number.
- The cubic polynomial $x^3 - (a + p)x^2 + (aq + r)x - as$ is divisible by $x - a$ if and only if $p = q$ and $r = s$.

Explorations

The sum of two consecutive numbers is not divisible by two, but the sum of three of them is divisible by three. When is the sum of a number of consecutive numbers divisible by that number? Generalize to consecutive terms from an arithmetic progression with specified common difference.

Given a closed continuous curve C drawn in the plane, and a point P in the plane, the winding number of C with respect to P is the number of times the curve winds around P. How does the winding number of C with respect to P change as P moves in the plane? What if the curve intersects itself more than once at a single point? Is there a way to calculate the winding number of C with respect to a given point P without actually traversing the curve C? Is there a connection between the number of crossings, the number of regions, and the winding number?

Imagine two coins of the same size lying on a table and just touching each other. One remains fixed, while the other rolls all the way round the first, as if they were both gears. How many times does the rolling coin rotate about its own centre (the rolling number of the configuration). Put another way, how many times does the design on the coin revolve? What if the coins are different sizes? What if the coin rolls around shapes other than a circle?

What connections if any are there between rolling numbers and winding numbers?

If the graph of $y = ax^2 + bx + c$ is translated parallel to the y axis at constant speed, at what speed do the roots move and in which direction? Try translating parallel to a general straight line, or even along a quadratic.

 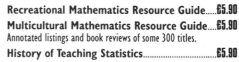

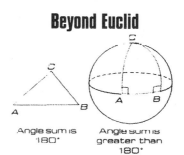

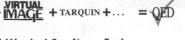 

FUN & GAMES with QED

BEAM + NCTM + Key Curriculum Press + + VIRTUAL IMAGE + TARQUIN + ... = QED

Puzzles and Games

FunMaths!
Junior FunMaths
BEAM Maths Calendar
Alice in Escherland
£6.90 each

FunMaths! is the acclaimed annual puzzle-pack full of fun and ideas for each day of the year. Eagerly awaited by 12,000 avid readers. Now joined by the BEAM Calendar and Junior FunMaths, a new any-year version, emphasising accessibility and ideas for kids aged 7 up.

"A goldmine ..." T.E.S.

FunFrench! £9.90
QED's first foray into language learning! Follows the tried and tested formula pioneered by FunMaths! (Cassette included.)

MathCounts (Photocopiables) £12.90
An American parallel to FunMaths! Full of Activities, Puzzles and Connections for 10-18 year-olds.

Recreational Mathematics £5.90
Multicultural Mathematics £5.90
Resource Guides, annotated listings and book reviews of 300 titles.

A Handbook of Cube-Assembly Puzzles £4.00
The definitive reference of puzzles using pentacubes etc. (Kevin Holmes).

Tilings and Patterns £34.00
The classic hardback by Grünbaum and Shephard (usually £84)

Maths and Humour

Full of quotes, cartoons, howlers etc., e.g. "How can an odd number like seven be made even?" Ans: "By taking away the S" (and other groans).

Mathematics and Humour £5.00
Not Strictly by the Numbers £9.90
Fifty Per Cent Proof: An Anthology of Mathematical Humour £5.90

New puzzle books from

Creative Puzzles of the World £19.95
Exploring Maths through Puzzles £14.95

One equals Zero and Other Mathematical Surprises: Paradoxes, Fallacies and Mind-bogglers £13.00
Illustrated coffee-table books for discerning puzzlists.

Puzzle-related books from the author of the acclaimed "Joy of Maths"

"Pappas has done it again!! These pages spill over with fun, ideas, puzzles and games from all over the world, past and present"
(from review)

The Joy of Mathematics, More Joy of Mathematics, The Magic of Mathematics, Fractals, Googols and other Mathematical Tales ..£10.90 each
Lots of links between maths, art, music, architecture, engineering and everyday life by the acclaimed author Theoni Pappas.

BEAM Board Games and other Devices

Board Games for the Nursery (2 sets of 5, A2 size) ....£19.50 each
Board Games for Infants (KS1-2) £15.00

NEW! colourful board-packs providing fun with learning, plus free Teacher Booklet. Available 1998/1999

Casting the Dice (1999) £7.50
Numbers in Your Head (1999) £7.50
Cards on the Table £7.50
Calculators in Their Hands £7.50
Four books of games for youngsters to enhance numeracy skills.

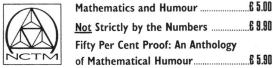

QED
1 Straylands Grove
York YO31 IEB
Tel: 01904 424242
Fax: 01904 424381
Email: qed@enterprise.net
School & Personal Orders: 0345 402275